WAR DEPARTMENT FIELD MANUAL

SERVICE OF THE PIECE
105-MM HOWITZER MOTOR CARRIAGE M7
PRIEST
FIELD MANUAL

RESTRICTED. DISSEMINATION OF RESTRICTED MATTER.—The information contained in restricted documents and the essential characteristics of restricted material may be given to any person known to be in the service of the United States and to persons of undoubted loyalty and discretion who are cooperating in Government work, but will not be communicated to the public or to the press except by authorized military public relations agencies. (See also par. 23b, AR 380-5, 15 Mar 1944.)

BY **WAR DEPARTMENT** • **10 JULY 1944**

©2013 Periscope Film LLC
All Rights Reserved
ISBN#978-1-940453-03-3
www.PeriscopeFilm.com

DISCLAIMER:

This manual is sold for historic research purposes only, as an entertainment. It contains obsolete information and is not intended to be used as part of an actual operation or maintenance training program. No book can substitute for proper training by an authorized instructor.

©2013 Periscope Film LLC
All Rights Reserved
ISBN#978-1-940453-03-3
www.PeriscopeFilm.com

WAR DEPARTMENT FIELD MANUAL

FM 6—74

This manual supersedes FM 17—63, 18 August 1942, including C1, 23 April 1943; C2, 13 September 1943; C3, 1 November 1943.

FIELD ARTILLERY

SERVICE OF THE PIECE

105-MM HOWITZER MOTOR CARRIAGE M7

WAR DEPARTMENT — 10 JULY 1944

RESTRICTED. *DISSEMINATION OF RESTRICTED MATTER.*—The Information contained in restricted documents and the essential characteristics of restricted material may be given to any person known to be in the service of the United States and to persons of undoubted loyalty and discretion who are cooperating in Government work, but will not be communicated to the public or to the press except by authorized military public relations agencies. (See also par. 236, AR 380-5, 15 Mar 1944.)

United States Government Printing Office
Washington 1944

WAR DEPARTMENT,
Washington 25, D. C., 10 July 1944.

FM 6-74, Field Artillery Field Manual, Service of the Piece, 105-mm Howitzer Motor Carriage M7, is published for the information and guidance of all concerned.

By Order of the Secretary of War:

G. C. MARSHALL,
Chief of Staff.

Official:

J. A. ULIO,
Major General,
The Adjutant General.

Distribution:

Base Comds (2); Island Comds (2); Def Comds (2); Sv C (continental and overseas) (2); Depts (2); Armies (2); Corps (2); Arm & Sv Boards (2); ROTC (1); D 17 (3); I Bn 6(70); IC 6(3).

I Bn 6: T/O & E 6-165;

IC 6: T/O & E 6-160-1.

TABLE OF CONTENTS

		Paragraphs	Page
SECTION	I. General	1-3	1
	II. Section composition and formations	4,5	4
	III. Posts, mounted and dismounted	6-9	6
	IV. Inspection and maintenance	10-13	10
	V. Prepare for action and march order	14-18	24
	VI. Indirect fire	19-33	34
	VII. Direct fire	34-50	55
	VIII. Additional information on service of the piece	51-61	71
	IX. Care and maintenance of material	62-68	76
	X. Destruction of equipment	69-75	87

Figure 1. 105-mm Howitzer on Motor Carriage M7.

RESTRICTED

WAR DEPARTMENT FIELD MANUAL

FIELD ARTILLERY

FM 6–74

SERVICE OF THE PIECE
105-MM HOWITZER MOTOR CARRIAGE M7*

Section I

GENERAL

---

1. **PURPOSE AND SCOPE.** This manual prescribes the duties to be performed in the service of the 105-mm howitzer motor carriage M7, by the personnel normally assigned to a howitzer section of the firing battery of an armored field artillery battery.

2. **REFERENCES.** a. Description, operation, functioning, and care of material. TM 9–731E; TM 9–325; TM 9–333; SNL C–21.

　b. Description and operation of fire control instruments. TM 6–220.

　c. Ammunition. TM 9–731E; TM 9–1900.

　d. Cleaning and preserving materials. TM 9–850.

　e. Vehicle maintenance and inspections. TM 9–731E; TM 9–2810; TM 9–333.

　f. Safety precautions in firing. AR 750–10; FM 6–40.

* For military terms not defined in this manual see TM 20–205.

g. **Gunnery.** FM 6-40.
h. **Duties in firing.** FM 6-40; FM 6-75.

3. DEFINITIONS AND TERMS. a. **Battery, armored field artillery battalion.** Tables of organization and equipment prescribe the personnel and materiel making up a battery, armored field artillery battalion. The battery consists of the battery headquarters, the firing battery, and the reconnaissance section.

b. **Firing battery.** The firing battery consists of the following:
 (1) Six howitzer sections.
 (2) One fire control section.
 (3) One ammunition section.

c. **Howitzer section.** A howitzer section is composed of one 105-mm howitzer on motor carriage M7, one ammunition trailer M10, and seven enlisted men.

d. **Fire control section.** The fire control section is composed of two half-tracks, the battery executive, the assistant executive, and the enlisted men who assist them.

e. **Ammunition section.** The ammunition section is composed of two half-tracks, two ammunition trailers, a chief of section, two drivers, and twelve ammunition handlers.

f. **Right or left.** The directions right or left are the right or left of the driver when he is in the driver's seat.

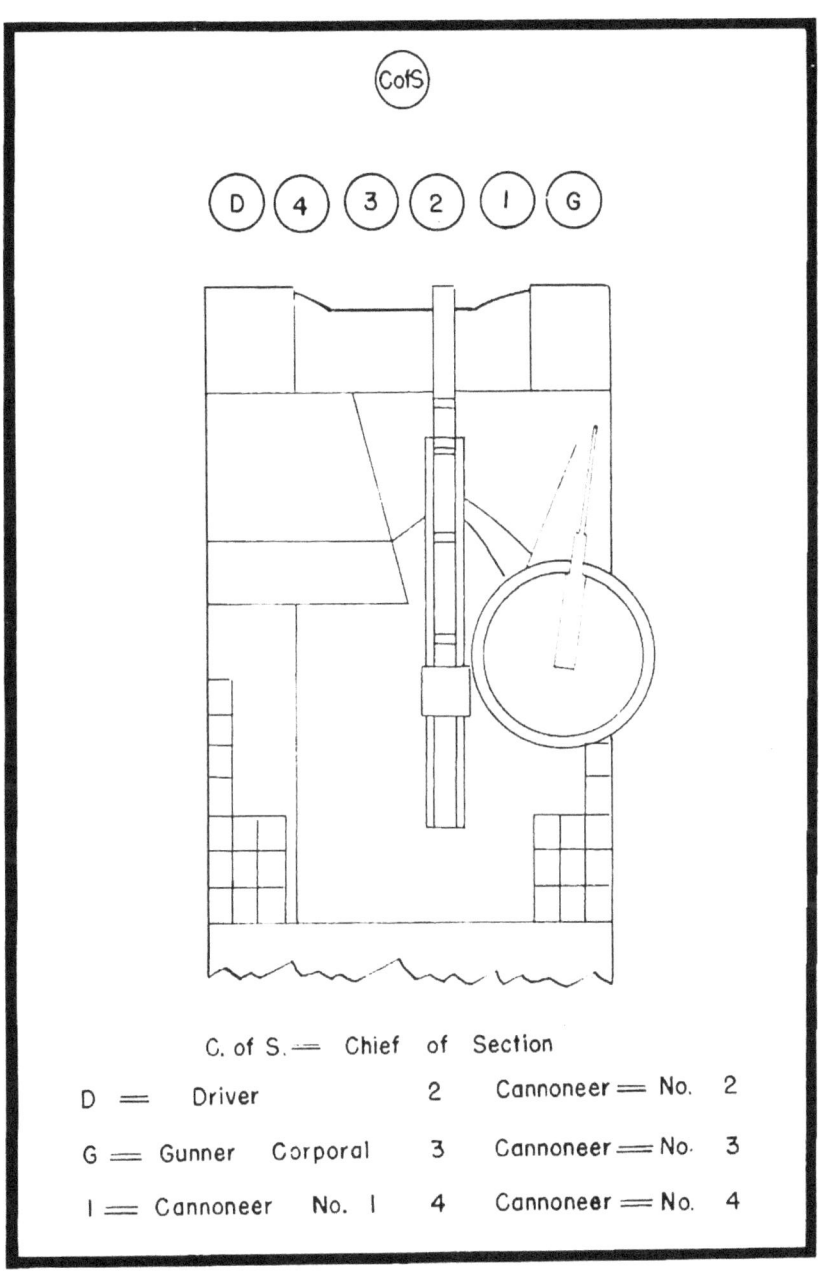

Figure 2. *Howitzer section, dismounted posts.*

Section II

SECTION COMPOSITION AND FORMATIONS

4. COMPOSITION. a. The personnel of the howitzer section consists of the following:
1 chief of section.
1 gunner.
4 cannoneers (Nos. 1, 2, 3, and 4).
1 driver.

b. **Relief cannoneers.** Ammunition handlers from the ammunition section act as relief cannoneers when so directed by the battery executive. They are assigned duties in the howitzer section by the chief of section.

5. FORMATIONS. a. **Dismounted.** For the first dismounted formation of the howitzer section for any drill or exercise, the order to FALL IN is preceded by AS HOWITZER SECTIONS. The place of formation may be stated in the command to fall in. For example: 1. AS HOWITZER SECTIONS, 2. IN FRONT OF YOUR PIECES, 3. FALL IN. The chief of section takes his post and supervises the formation. The gunner repeats the command FALL IN and takes his post, faced in the proper direction, at the point where the right of the section is to rest. The remainder of the section moves at double time and falls in at the gunner's left as shown in figure 2.

b. **Call off.** (1) The chief of section commands CALL OFF. The gunner calls off GUNNER; the cannoneer on the left of the gunner calls off ONE; the cannoneer on the left of No. 1, TWO; and so on, ending with the driver who calls off DRIVER.

(2) After having called off, if a subsequent formation is ordered, the cannoneers fall in at once in their proper order.

c. **Change posts.** In order to train each member of the section in all duties connected with the service of the piece, the posts of individual cannoneers should be changed frequently. The cannoneers being formed in their dismounted posts, the command is 1. CHANGE POSTS, 2. MARCH. At this command the gunner quickly takes the post of No. 1, No. 1 of No. 2, and so on, ending with the driver, who takes the post of the gunner.

Section III

POSTS, MOUNTED AND DISMOUNTED

6. POSTS OF THE HOWITZER SECTION. a. Dismounted. The posts of the howitzer section dismounted are shown in figure 2.

b. Mounted. The posts of the howitzer section mounted are shown in figure 3.

c. Prepared for action. The posts of the howitzer section prepared for action are shown in figure 4.

7. TO MOUNT THE HOWITZER SECTION. Being at dismounted posts or dismounted, the section is mounted at the command or signal MOUNT. At the command or signal, the section mounts as follows:

a. The driver, gunner, and No. 2 mount over the left side of the motor carriage and take their mounted posts.

b. No. 1, No. 3, and No. 4 mount over the right side of the motor carriage and take their mounted posts.

c. After all other members of the section have mounted, the chief of section mounts over the left side of the motor carriage and takes his mounted post.

d. The command MOUNT may be preceded by the command PREPARE TO MOUNT.

8. TO DISMOUNT THE HOWITZER SECTION. The section, being at mounted posts or mounted, is dismounted at the command or signal 1. PREPARE TO DISMOUNT, 2. DISMOUNT. At the preparatory com-

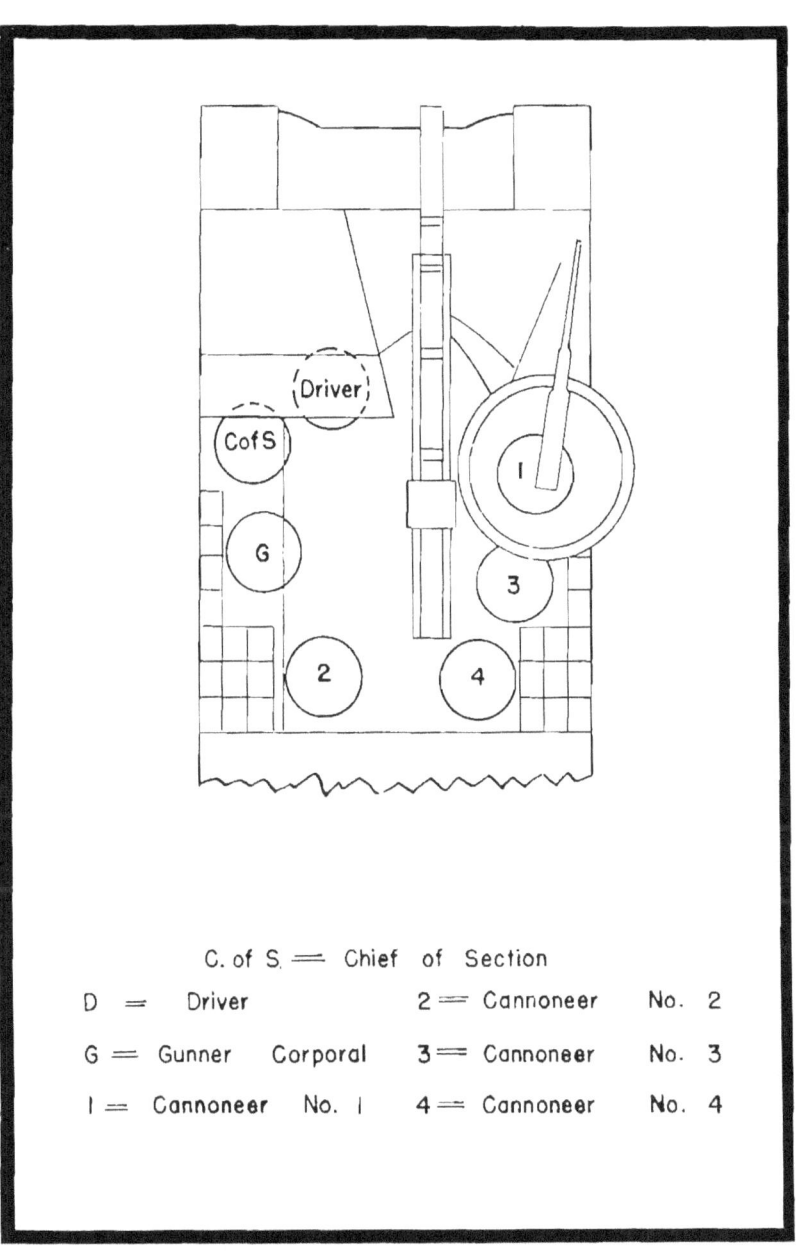

Figure 3. Howitzer section, mounted posts.

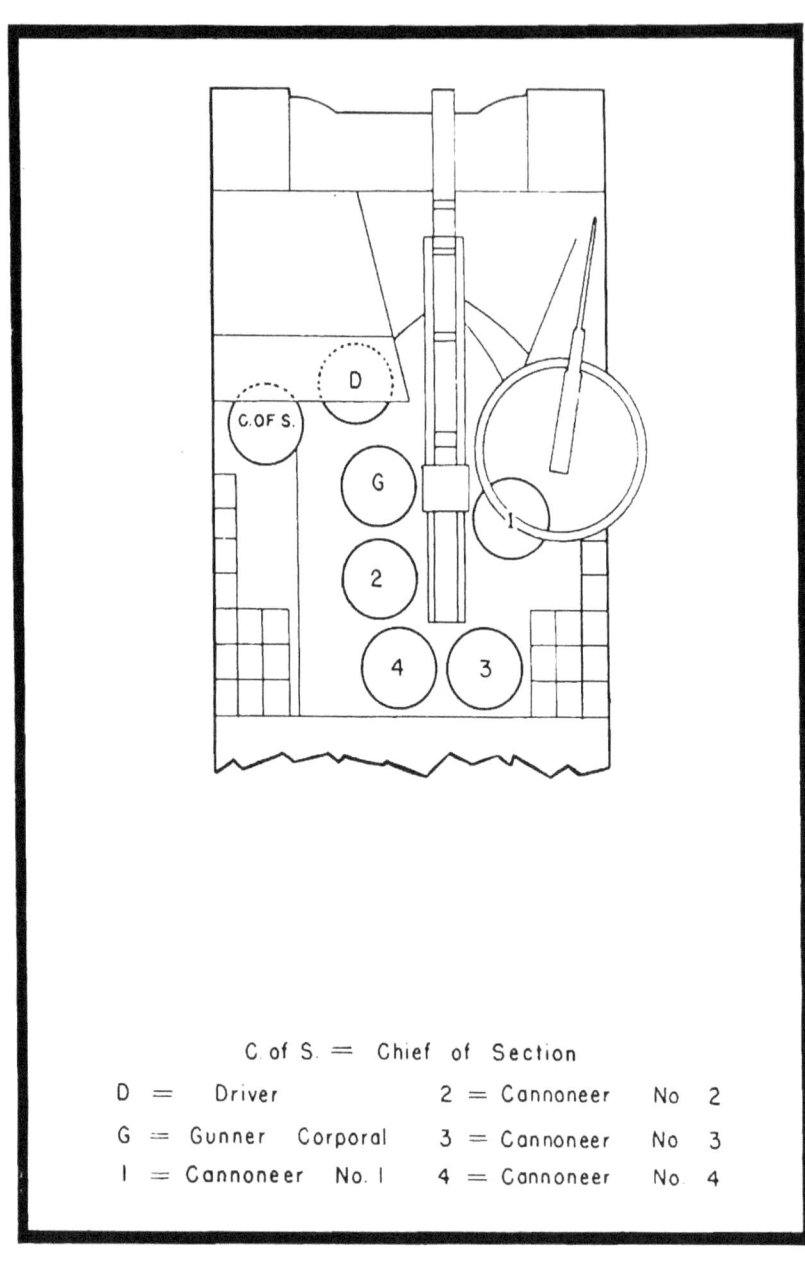

Figure 4. Howitzer section, action posts.

mand members of the section stow all vehicular equipment. At the command DISMOUNT, members of the section dismount in reverse of the order prescribed in paragraph 7 (TO MOUNT THE HOWITZER SECTION) and take their dismounted posts (figure 2).

9. TO POST THE SECTION. The section being at rest or dismounted from the vehicle and not in any formation, the command is 1. CANNONEERS, 2. POSTS. Each gunner repeats the command POSTS, and the driver and cannoneers move at double time to the action posts shown in figure 4. If the section is dismounted, they mount as shown in paragraph 7, and then take action posts.

Section IV

INSPECTION AND MAINTENANCE

10. INSPECTION BEFORE OPERATION. The inspection performed before operation is a final check on materiel before going into combat. It is performed in the bivouac or motor park. At the completion of this inspection, and after all deficiencies have been corrected, the motor carriage and armament are ready to go into action. The section being at dismounted posts, the chief of section commands PREPARE FOR INSPECTION, and inspects the individual equipment of members of the section. He then commands INSPECT EQUIPMENT, and the section proceeds as follows:

a. **Chief of Section.** (1) Examines recoil cylinder for leakage of oil.

(2) Verifies the fact that the recoil mechanism contains the proper amount of oil. He orders the Gunner and No. 1 to service it when necessary.

(3) Inspects suspension system and tracks.

(4) Inspects ammunition trailer.

(5) Inspects all vehicular equipment.

(6) Inspects ammunition.

(7) Checks telephone communication system.

(8) Performs "End-for-End" test on gunner's quadrant.

(9) Supervises detailed inspection by all members of the section.

(10) Verifies presence of firing tables, gun book, and manuals.

(11) Checks emergency supply of rations and water.
(12) Inspects flag set.
(13) Designates cannoneer to inspect tools, spare parts and accessories, and supervises this inspection.
(14) Takes mounted post.
(15) Reports (SO AND SO) SECTION IN ORDER to the battery executive.

b. Driver. (1) Inspects ground beneath motor carriage for evidence of fuel leaks, oil leaks, or coolant leaks.

(2) Inspects engine and compartment for presence of fumes and evidence of fuel or oil leaks.

°(3) Turns fuel filter handle one complete turn in either direction.

(4) Checks transmission and final drive oil level.

(5) Mounts and makes visual check of clutch release bearing. (Not required on M4A3 chassis)

(6) Turns on main battery switch.
(7) Checks fuel supply and gauges.
(8) Opens inside fuel shut-off valves.
(9) Checks engine oil level.
(10) Checks clutch pedal free travel.
(11) Sees that gear shift lever latch is operating and that the lever can be put into each position.
(12) Tests action of parking brakes.
(13) Tests operation of steering levers.
(14) Tests siren switch and fuel cut-off switch.
(15) Primes engine.
(16) Starts and warms up engine.
(17) Notes operation of all instruments on panel.
(18) Performs RPM drop check.
(19) Checks operation of all lights, assisted by No. 4.
(20) Inspects windshield and protectoscope.
(21) Moves motor carriage so entire track can be inspected by No. 4.

°On M4A3 chassis omit and substitute:
   (3) Checks coolant level.

(22) Reports DRIVER READY.

c. Gunner. (1) Unfastens and folds back breech end of howitzer cover, assisted by No. 1.

(2) Moves piece to assist No. 2 in disengaging howitzer traveling lock.

(3) Tests operation of elevating handwheel on left side of howitzer.

(4) Tests operation of traveling mechanism and traverses piece to right as far as possible so driver can check transmission and final drive oil level.

(5) Removes elbow telescope from case and passes it to No. 1.

(6) Removes panoramic telescope from case, seats it in mount, and checks it for cleanliness.

(7) Boresights panoramic telescope.

(8) Tests night lighting device on panoramic telescope.

(9) Replaces panoramic telescope in case.

(10) Receives elbow telescope from No. 1 and replaces it in case.

(11) Inspects leveling devices on telescope mount.

(12) Services recoil cylinder, assisted by No. 1, when so ordered by chief of section.

(13) Moves piece to assist No. 2 in engaging howitzer traveling lock.

(14) Fastens breech end of howitzer cover, assisted by No. 1.

(15) Takes mounted post.

(16) Reports GUNNER READY.

d. Cannoneer No. 1. (1) Assists gunner in unfastening breech end of howitzer cover.

(2) Tests operation of elevating mechanism.

(3) Tests operation of breech mechanism and inspects breechblock, chamber, and bore.

(4) Receives elbow telescope from gunner.

(5) Boresights elbow telescope.

(6) Tests night lighting devices.

(7) Checks range quadrant by gunner's quadrant.
(8) Passes elbow telescope to gunner.
(9) Removes cover from machine gun.
(10) Inspects and half-loads machine gun.
(11) Assists gunner in servicing recoil cylinder, when so ordered by chief of section.
(12) Assists gunner in fastening breech end of howitzer cover.
(13) Takes mounted post.
(14) Reports NUMBER ONE READY.

e. Cannoneer No. 2. (1) Disengages howitzer traveling lock.
(2) Removes traveling lock bracket, assisted by No. 3.
(3) Removes breech end of howitzer cover, assisted by No. 3, and places it on rear deck of motor carriage.
(4) Removes firing lock or places rear bore sight in tube.
(5) Checks grenades.
(6) Checks first aid kit.
(7) Checks fire extinguishers.
(8) Checks decontaminating devices.
(9) Replaces firing lock, or removes rear bore sight.
(10) Replaces traveling lock bracket, assisted by No. 3.
(11) Engages howitzer traveling lock.
(12) Replaces breech end of howitzer cover, assisted by No. 3.
(13) Takes mounted post.
(14) Reports NUMBER TWO READY.

f. Cannoneer No. 3. (1) Assists No. 2 in removing traveling lock bracket.
(2) Places traveling lock bracket on floor of motor carriage.
(3) Assists No. 2 in removing breech end of howitzer cover.
(4) Inspects fuze setter.

(5) Checks howitzer ammunition supply in motor carriage.

(6) Removes muzzle end of howitzer cover.

(7) Attaches cross hairs for boresighting to muzzle of howitzer.

(8) Sets out test target, when so ordered by gunner.

(9) Checks ammunition supply in trailer.

(10) Checks loading and strapping of ammunition in trailer.

(11) Makes certain that trailer tail gate is closed and properly fastened.

(12) Checks trailer brakes.

(13) Inspects trailer tires for inflation and casing breaks.

(14) Inspects trailer tools and equipment and fastens fuze box lid.

(15) Checks trailer connection with motor carriage.

(16) Checks trailer stand to see that it is in "UP" position and properly fastened.

(17) Brings in test target.

(18) Removes cross hairs from muzzle of howitzer.

(19) Puts cover on muzzle of howitzer.

(20) Assists No. 2 in replacing traveling lock bracket.

(21) Assists No. 2 in replacing breech end of howitzer cover.

(22) Takes mounted post.

(23) Reports NUMBER THREE READY.

g. Cannoneer No. 4. (1) Opens engine compartment doors.

°(2) Turns over engine with 50 revolutions of hand crank.

(3) Checks air cleaners for security.

(4) Checks auxiliary fuel tanks for security.

(5) Assists driver in checking all lights on motor carriage and ammunition trailer while engine warms up.

°Omit with M4A3 chassis.

(6) Inspects tracks for tension.

(7) Instructs driver, by hand signals, to move motor carriage so entire track can be inspected.

(8) Inspects condition of tracks.

(9) Inspects sprockets.

(10) Inspects bogie wheels.

(11) Inspects bogie assemblies.

(12) Inspects support rollers.

(13) Inspects idlers.

(14) Inspects for oil and fuel leaks under hull and in engine compartment with engine running.

(15) Closes engine compartment doors.

(16) Takes mounted post.

(17) Reports NUMBER FOUR READY.

11. INSPECTION DURING OPERATION. The inspection performed during operation is a constant check on the operation of the motor carriage and the security of all stowed equipment, to insure that all materiel goes into combat in the best possible condition. There is no command for this inspection, as it is carried on constantly. The responsibilities and duties are as follows:

a. Chief of section. (1) Be alert to any unusual noises or conditions.

(2) Assists driver to avoid obstacles that would cause injury to motor carriage or members of the section.

b. Driver. (1) Checks all instruments carefully.

(2) Operates engine at 1800 to 2100 rpm.

(3) Keeps foot off clutch pedal except when using clutch.

(4) Keeps hands off steering levers except when using them.

(5) Listens for unusual noises, looks for smoke or steam and notes unusual odors.

c. Gunner. (1) Checks security of the howitzer traveling lock.

(2) Checks gun covers, if on.

d. Cannoneer No. 1. (1) Checks security of caliber .50 machine gun.

(2) Air sentinel to front.

e. Cannoneer No. 2. Checks security of 105-mm ammunition.

f. Cannoneer No. 3. (1) Checks security of stowed equipment in general.

(2) Air sentinel to rear.

(3) Notes operation of trailer.

g. Cannoneer No. 4. (1) Checks for unusual engine noises.

(2) Takes driver's place when necessary.

12. INSPECTION AT HALTS. The inspection at the halt is made to insure that the motor carriage and armament are in good condition. This inspection allows the members of the section to check the things that cannot be checked during operation and is performed as soon as a halt in marching is made. The duties of the members of the section are as follows:

a. Chief of section. (1) Checks general condition of motor carriage, trailer, and armament.

(2) Supervises inspection by members of the section.

(3) Makes general check of tracks, sprockets, and suspension system.

(4) Commands REPORT.

(5) Reports NUMBER (SO AND SO) IN ORDER, to the battery executive.

b. Driver. (1) Performs RPM drop check.

(2) Idles engine at 800 rpm three to five minutes, noting engine operation.

(3) Notes operation of all instruments on panel.

(4) Notes operation of engine controls.

(5) Moves motor carriage so entire track can be inspected by No. 4.

(6) Checks windshield and protectoscope.
(7) Stops engine.
(8) Checks fuel and oil supply, replenishing if necessary. In case of M4A3 chassis, checks coolant level.
(9) Reports DRIVER READY on command from chief of section.

c. Gunner. (1) Checks howitzer traveling lock for security.
(2) Examines tape cover on muzzle. If cover is broken, instructs No. 3 to swab bore. Puts on new cover.
(3) Boresights panoramic telescope if time permits.
(4) Reports GUNNER READY on command from chief of section.

d. Cannoneer No. 1. (1) Checks caliber .50 machine gun and ammunition.
(2) Boresights elbow telescope if time permits.
(3) Reports NUMBER ONE READY on command from chief of section.

e. Cannoneer No. 2. (1) Checks 105-mm ammunition in howitzer for security.
(2) Removes and replaces firing lock or installs and removes rear bore sight as required.
(3) Reports NUMBER TWO READY on command from chief of section.

f. Cannoneer No. 3. (1) Checks general mechanical condition of trailer.
(2) Checks ammunition in trailer for security.
(3) Attaches and removes cross hairs, as required for boresighting.
(4) Reports NUMBER THREE READY on command from chief of section.

g. Cannoneer No. 4. (1) Opens engine compartment doors and inspects engine compartment.
(2) Checks air cleaners for security.
(3) Inspects tracks and suspension system, removing rocks, mud, etc.

(4) Instructs driver, by hand signals, to move motor carriage so entire track can be inspected.

(5) Checks temperature of differential and final drives.

(6) Reports NUMBER FOUR READY on command from chief of section.

13. MAINTENANCE AFTER OPERATION. The maintenance operations performed after operation are to put the motor carriage and armament back into proper condition to go into combat or to determine what must be done by the higher echelons of maintenance. These operations are performed in a bivouac area, motor park, or combat position. With the section at dismounted posts, the chief of section commands INSPECT EQUIPMENT, and the section proceeds as follows:

a. Chief of section. (1) Makes general check of motor carriage, trailer, and armament.

(2) Examines recoil cylinder for leakage of oil.

(3) Examines oil index to see that recoil cylinder has "full reserve". (If recoil cylinder has a lack of oil reserve, orders gunner and No. 1 to service it.)

(4) Inspects suspension system and tracks.

(5) Mounts and inspects all vehicular armament.

(6) Tests telephone, wire, and reel.

(7) Cleans gunner's quadrant and performs "End-for-End" test.

(8) Inspects ammunition.

(9) Inspects grenades.

(10) Inspects first aid kit.

(11) Inspects decontaminating equipment.

(12) Inspects fire extinguishers.

(13) Inspects firing tables and manuals.

(14) Posts gun book.

(15) Inspects flag set.

(16) Designates cannoneer to inspect tools, spare parts and accessories, and supervises this inspection.

(17) Reports to battery executive requirements for:
Fuel, oil, and lubricating materials.
Ammunition and grenades.
Maintenance and repairs.
Replacement of personnel.
First aid kit.
Decontaminating equipment.
Fire extinguishers.
Rations and water.
(18) Takes dismounted post.
(19) Reports (SO AND SO) SECTION IN ORDER, to battery executive.

b. Driver. (1) Performs RPM drop check.

(2) Idles engine at 800 rpm three to five minutes, noting engine operation.

(3) Notes operation of engine controls.

(4) Notes operation of all instruments on panel.

(5) Tests operation of steering levers.

(6) Tests operation of all lights assisted by No. 4.

(7) Tests operation of siren.

(8) Checks protectoscope and windshield.

(9) Moves motor carriage so entire track can be inspected by No. 4.

(10) Checks clutch pedal free travel.

(11) Stops engine, using fuel cut-off switch.

(12) Turns off ignition switch.

(13) Turns off main battery switch.

(14) Closes inside fuel shut-off valves.

(15) Checks fire extinguishers, assisted by No. 4.

*(16) Checks clearance, and lubricates clutch release bearing.

(17) Checks engine oil level.

(18) Checks engine electrical wiring and accessories.

(19) Checks differential lubricant level.

*On M4A3 chassis, omit and substitute:
(16) Checks coolant.

(20) Checks transmission and final drive oil level.
(21) Reports to chief of section requirements for:
Fuel.
Lubricants.
Maintenance.
(22) Refills fuel tanks, assisted by No. 4.
(23) Takes dismounted post.
(24) Reports DRIVER READY.

c. Gunner. (1) Unfastens and folds back breech end of howitzer cover, assisted by No. 1.

(2) Moves piece to assist No. 2 in disengaging howitzer traveling lock.

(3) Services recoil cylinder assisted by No. 1 when so ordered by chief of section.

(4) Tests operation of traversing mechanism.

(5) Cleans traversing mechanism.

(6) Tests operation of elevating handwheel on left side of howitzer.

(7) Cleans and inspects telescope mount.

(8) Removes elbow telescope from case and passes it to No. 1.

(9) Removes panoramic telescope from case and cleans it.

(10) Seats panoramic telescope in mount.

(11) Boresights panoramic telescope. (If test target is used, orders No. 3 to set up target.)

(12) Cleans and tests night lighting device.

(13) Replaces panoramic telescope in case.

(14) Receives elbow telescope from No. 1 and places it in case.

(15) Moves piece to assist No. 2 in engaging howitzer traveling lock.

(16) Reports maintenance requirements to chief of section.

(17) Fastens breech end of howitzer cover, assisted by No. 1.

(18) Takes dismounted post.
(19) Reports GUNNER READY.

d. Cannoneer No. 1. (1) Assists gunner in unfastening and folding back breech end of howitzer cover.

(2) Assists gunner in servicing recoil cylinder when so ordered by chief of section.

(3) Tests operation of elevating mechanism.
(4) Cleans elevating mechanism.
(5) Cleans elbow telescope mount.
(6) Receives elbow telescope from gunner and cleans it.
(7) Seats elbow telescope in mount.
(8) Boresights elbow telescope.
(9) Cleans and tests night lighting devices.
(10) Passes elbow telescope to gunner.
(11) Cleans and inspects range quadrant.
(12) Checks range quadrant by gunner's quadrant.
(13) Cleans and tests firing mechanism.
(14) Unloads and dismounts machine gun.
(15) Cleans machine gun and replaces it on mount.
(16) Checks ammunition for machine gun.
(17) Reports maintenance and ammunition requirements for machine gun to chief of section.
(18) Checks spare parts for machine gun.
(19) Places cover on machine gun.
(20) Assists gunner in fastening breech end of howitzer cover.
(21) Takes dismounted post.
(22) Reports NUMBER ONE READY.

e. Cannoneer No. 2. (1) Disengages howitzer traveling lock.

(2) Removes traveling lock bracket assisted by No. 3.
(3) Removes breech end of howitzer cover, assisted by No. 3, and places it on rear deck of motor carriage.
(4) Cleans breechblock, assisted by No. 3.
(5) Cleans bore of howitzer, assisted by No. 3.

(6) Cleans recoil slides, assisted by No. 3.

(7) Removes firing lock, or places rear bore sight in tube.

(8) Replaces firing lock or removes bore sight from tube.

(9) Refills ammunition racks in howitzer, assisted by No. 3.

(10) Replaces traveling lock bracket, assisted by No. 3.

(11) Engages howitzer traveling lock.

(12) Replaces breech end of howitzer cover, assisted by No. 3.

(13) Takes dismounted post.

(14) Reports NUMBER TWO READY.

f. Cannoneer No. 3. (1) Assists No. 2 in removing traveling lock bracket.

(2) Places traveling lock bracket on floor of motor carriage.

(3) Assists No. 2 in removing breech end of howitzer cover.

(4) Assists No. 2 in cleaning breechblock.

(5) Assists No. 2 in cleaning bore of howitzer.

(6) Assists No. 2 in cleaning recoil slides.

(7) Attaches cross hairs for boresighting to muzzle of howitzer.

(8) Sets out test target when so ordered by the gunner.

(9) Checks trailer welds for cracks.

(10) Checks trailer tires for cuts, and inflates to 40 pounds pressure.

(11) Checks trailer wheel nuts for tightness.

(12) Checks all trailer body bolts and nuts, making certain they are tight.

(13) Assists No. 2 in refilling ammunition racks in howitzer.

(14) Brings in test target.

(15) Removes cross hairs from the muzzle.

(16) Puts canvas cover on muzzle.

(17) Assists No. 2 in replacing traveling lock bracket.

(18) Assists No. 2 in replacing breech end of howitzer cover.

(19) Takes dismounted post.

(20) Reports NUMBER THREE READY.

g. **Cannoneer No. 4.** (1) Opens engine compartment doors.

(2) Notes engine operation and checks for fuel and oil leaks while engine is running.

(3) Checks tracks for tension.

(4) Checks tracks for condition.

(5) Instructs driver, by hand signals, to move motor carriage so entire track can be inspected.

(6) Checks sprockets.

(7) Checks bogie wheels.

(8) Checks bogie assemblies.

(9) Checks support rollers.

(10) Checks idlers.

(11) Inspects for oil leaks under hull.

(12) Assists driver in checking all lights while engine cools.

(13) Assists driver in checking fire extinguishers.

(14) Cleans pintle hook and checks operation.

(15) Services oil bath air cleaners as required.

(16) Inspects automatic oil filter.

(17) Closes engine compartment doors.

(18) Assists driver in refilling fuel tanks.

(19) Takes dismounted post.

(20) Reports NUMBER FOUR READY.

NOTE: When available, ammunition handlers are assigned to each howitzer section to assist with cleaning and maintenance of equipment.

Section V

PREPARE FOR ACTION
AND MARCH ORDER

14. AMMUNITION TRAILER. a. General. When a howitzer section prepares for action, if possible, the ammunition trailer is uncoupled and left a short distance to the rear of the final firing position of the motor carriage. The trailer must never be allowed to handicap the movements of the motor carriage during firing.

b. Uncouple trailer. To uncouple the ammunition trailer, the chief of section signals the driver to stop the motor carriage and then gives the command UNCOUPLE TRAILER. Cannoneers Nos. 2, 3, and 4 dismount and Nos. 3 and 4 set the hand parking brakes. No. 2 lowers the stand, making sure that it is in the full "DOWN" position. No. 2 then unlocks the pintle hook at the rear of the motor carriage, while Nos. 3 and 4 lift the lunette out. While Nos. 3 and 4 hold the trailer handles, No. 2 signals the chief of section to have the motor carriage moved forward. The chief of section, by touch signals to the driver, controls the movement of the motor carriage into the firing position. Nos. 3 and 4 set the trailer down, and all cannoneers proceed at double time to the motor carriage.

c. Couple trailer. At the command COUPLE TRAILER, the driver starts the engine if it is not running and cannoneers Nos. 2, 3, and 4 dismount and proceed at double time to the trailer. The chief of section, by touch signals to the driver, controls the move-

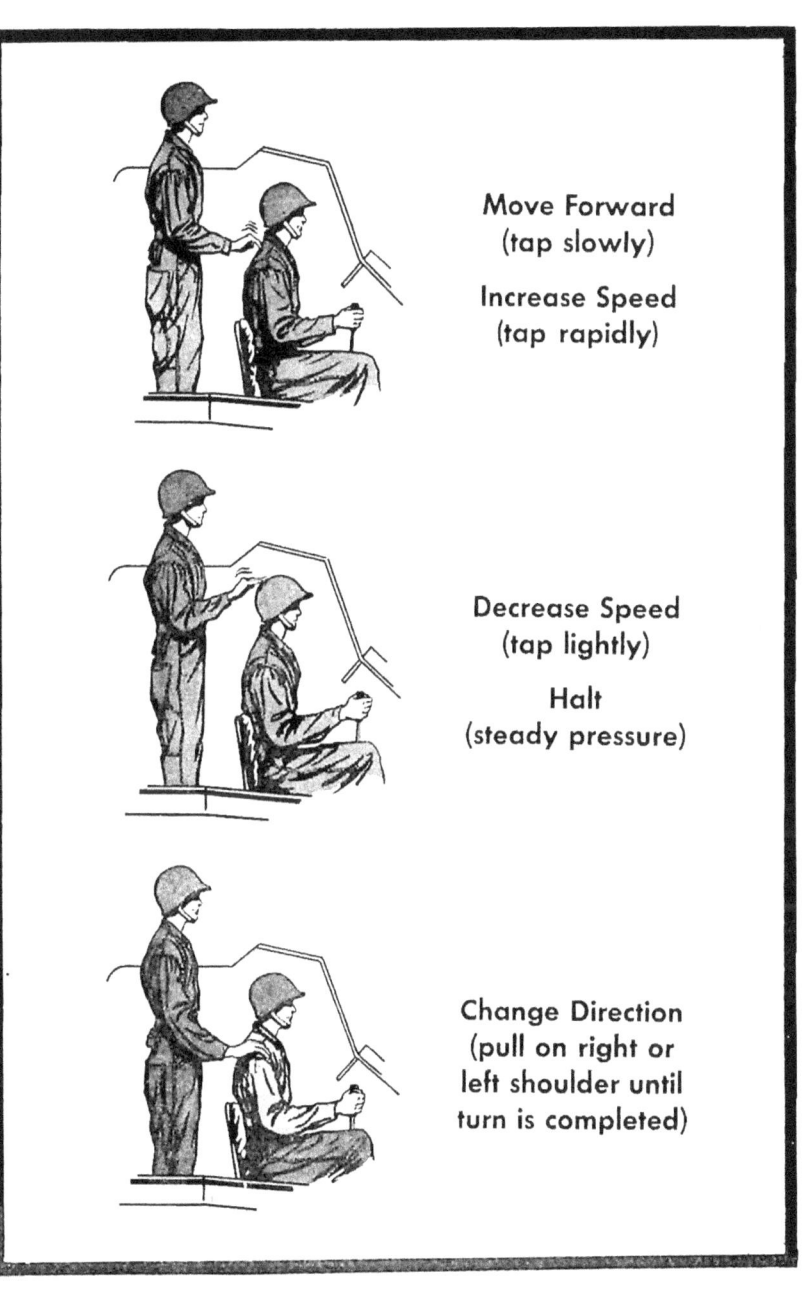

Figure 5. Touch signals.

ment of the motor carriage back to the position of the trailer. While No. 2 opens the pintle hook at the rear of the motor carriage, Nos. 3 and 4 hold the trailer. No. 2 signals the chief of section to have the motor carriage back up to the lunette. Nos. 3 and 4 place the lunette in the pintle hook and No. 2 locks it. Nos. 3 and 4 release the hand parking brakes while No. 2 raises the trailer support stand. The cannoneers then return to their mounted posts in the motor carriage.

15. TOUCH SIGNALS. The following are the signals used by the chief of section and the gunner to indicate to the driver the proper movement of the motor carriage. Some of them are illustrated in figure 5.

   a. Move forward. Tap slowly between shoulder blades.

   b. Increase speed. Tap rapidly between shoulder blades.

   c. Halt. Steady pressure on top of the head.

   d. Decrease speed. Tap lightly on top of the head.

   e. Move backward. Tug on back of collar.

   f. Change direction. Pull on right or left shoulder until turn is completed.

   g. Sound siren. Run thumb rapidly upward between shoulder blades.

16. HAND SIGNALS. The following are the signals used to indicate to the driver the proper movement of the motor carriage. They are given from a dismounted position, facing the driver, and are illustrated in figures 6 and 7.

   a. Start engine. Describe a cranking motion in front of body.

   b. Stop engine. Cross forearms repeatedly in front of body.

   c. Move forward or backward. Motion with repeated movements of both hands in front of face, palms facing

Figure 6. Hand signals.

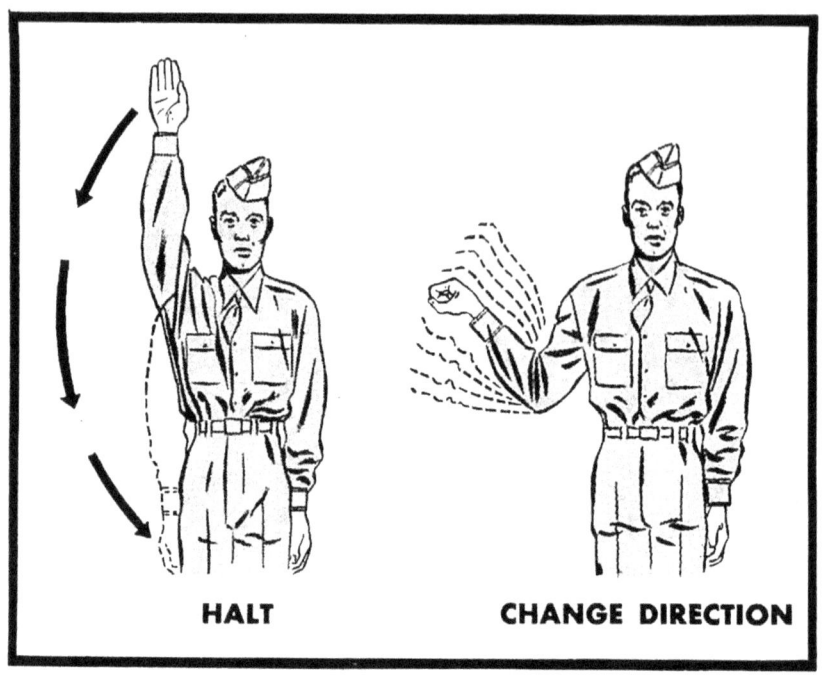

Figure 7. Hand signals.

the direction of travel. Speed of movement is controlled by increasing rate of motions.

d. Halt. Clasp hands in front of face. Warning for halt is given by holding arms to the front, fingers extended and joined, palms turned inward. Hands are moved together to approximate the amount of movement still to be made.

e. Change direction. Clench and slightly raise the fist on the side to which the turn is to be made (side on which brake is applied).

17. PREPARE FOR ACTION. When the firing battery arrives in a previously selected firing position or receives a fire mission while on the march, Prepare for Action will be executed. At the command or signal

PREPARE FOR ACTION, if time permits, the trailer is uncoupled as explained in paragraph **14b**. The section then proceeds as follows:

a. Chief of section. (1) Commands PREPARE FOR ACTION.

(2) Gives signals to driver to move motor carriage into firing position.

(3) Supervises work of all members of the section.

(4) Prepares telephone for use and checks communication.

(5) Commands REPORT.

(6) Reports NUMBER (SO AND SO) READY to battery executive.

b. Driver. (1) Remains at his post.

(2) Keeps engine running.

(3) Reports DRIVER READY on command from chief of section.

c. Gunner. (1) Unfastens and folds back breech end of howitzer cover, assisted by No. 1.

(2) Moves piece to assist No. 2 in disengaging howitzer traveling lock.

(3) Removes elbow telescope from case and passes it to No. 1.

(4) Removes panoramic telescope from case and seats it in mount.

(5) Sets elevation indexes, azimuth scale, and azimuth micrometer of panoramic telescope at zero.

(6) Opens telescope mount level covers.

(7) Centers cross and longitudinal level bubbles.

(8) Boresights panoramic telescope, if time permits.

(9) Reports GUNNER READY on command from chief of section.

d. Cannoneer No. 1. (1) Assists gunner in unfastening and folding back breech end of howitzer cover.

(2) Opens breech and inspects chamber.

(3) Receives elbow telescope from gunner and seats it in mount.

(4) Sets angle of site at 300.

(5) Sets elevation at 200.

(6) Opens range quadrant level covers.

(7) Centers cross and longitudinal level bubbles.

(8) Boresights elbow telescope, if time permits.

(9) Reports NUMBER ONE READY on command from chief of section.

e. Cannoneer No. 2. (1) Disengages howitzer traveling lock.

(2) Removes traveling lock bracket and places it on floor of motor carriage, assisted by No. 3.

(3) Removes breech end of howitzer cover, assisted by No. 3, and places it on rear deck of motor carriage.

(4) Assembles aiming posts and rammer staff, and places them on rear deck of motor carriage, or sets out aiming posts as ordered.

(5) Removes and replaces firing lock or installs and removes rear bore sight as required for boresighting.

(6) Reports NUMBER TWO READY on command from chief of section.

f. Cannoneer No. 3. (1) Assists No. 2 in removing traveling lock bracket and placing it on floor of motor carriage.

(2) Assists No. 2 in removing breech end of howitzer cover.

(3) Spreads tarpaulin on rear deck of motor carriage, assisted by No. 4.

(4) Places ammunition trough on tarpaulin.

(5) Places fuze setter in convenient position.

(6) Prepares ammunition for firing, assisted by No. 4.

(7) Attaches cross hairs for boresighting to muzzle of howitzer.

(8) Reports NUMBER THREE READY on command from chief of section.

g. **Cannoneer No. 4.** (1) Lays wire for intrabattery communication system.

(2) Assists No. 3 in spreading ammunition tarpaulin on rear deck of motor carriage.

(3) Assists No. 3 in preparing ammunition for firing.

(4) Reports NUMBER FOUR READY on command from chief of section.

18. MARCH ORDER. When a displacement of some distance is to be made, March Order will be executed. If the trailer is not coupled to the motor carriage at the completion of the duties listed below, and if time permits, the trailer is coupled to the motor carriage as shown in paragraph 14c. At the command or signal MARCH ORDER, the section proceeds as follows:

a. Chief of section. (1) Commands MARCH ORDER.

(2) Inspects chamber to see that piece is not loaded.

(3) Supervises work of all members of the section.

(4) Tests traveling lock bracket to make sure piece is locked in traveling position.

(5) Replaces telephone head and chest set.

(6) Takes mounted post.

(7) Commands REPORT.

(8) Reports NUMBER (SO AND SO) IN ORDER to battery executive.

b. Driver. (1) Remains at his post.

(2) Starts engine, if it is not running.

(3) Reports DRIVER READY on command from chief of section.

c. Gunner. (1) Moves piece to assist No. 2 in engaging howitzer traveling lock.

(2) Sets elevation indexes, azimuth scale, and azimuth micrometer of panoramic telescope at zero.

(3) Returns panoramic telescope to case.

(4) Receives elbow telescope from No. 1 and returns it to case.

(5) Closes telescope mount level covers.

(6) Fastens breech end of howitzer cover, assisted by No. 1.

(7) Takes mounted post.

(8) Reports GUNNER READY on command from chief of section.

d. Cannoneer No. 1. (1) Inspects chamber to see that piece is not loaded.

(2) Closes breech after inspection of chamber by chief of section.

(3) Removes elbow telescope from mount and passes it to gunner.

(4) Sets angle of site at 300.

(5) Replaces Zone VII range drum, if necessary.

(6) Closes range quadrant lever covers.

(7) Assists gunner in fastening breech end of howitzer cover.

(8) Takes mounted post.

(9) Reports NUMBER ONE READY on command from chief of section.

e. Cannoneer No. 2. (1) Replaces traveling lock bracket assisted by No. 3.

(2) Engages howitzer traveling lock.

(3) Replaces breech end of howitzer cover, assisted by No. 3.

(4) Recovers, disassembles, and secures in traveling position the aiming posts and rammer staff.

(5) Checks ammunition and refills racks in motor carriage if time permits.

(6) Takes mounted post.

(7) Reports NUMBER TWO READY on command from chief of section.

f. Cannoneer No. 3. (1) Assists No. 2 in replacing traveling lock bracket.

(2) Returns ammunition trough to traveling position.

(3) Stows fuze setter.

(4) Folds ammunition tarpaulin, assisted by No. 4.

(5) Assists No. 2 in replacing breech end of howitzer cover.

(6) Dismounts and puts cover on muzzle of howitzer.

(7) Checks ammunition and, if time permits, assists No. 2 in refilling ammunition racks in motor carriage.

(8) Takes dismounted post.

(9) Reports NUMBER THREE READY on command from chief of section.

g. Cannoneer No. 4. (1) Rewinds telephone wire.

(2) Assists No. 3 in folding and securing ammunition tarpaulin.

(3) Assists No. 2 and No. 3 in refilling ammunition racks in motor carriage with ammunition from trailer, if time permits.

(4) Takes mounted post.

(5) Reports NUMBER FOUR READY on command from chief of section.

Section VI

INDIRECT FIRE

19. GENERAL. The principal duties in firing for each member of the section are listed below. For complete lists and detailed descriptions of the duties of each member of the section see paragraphs **20-33** inclusive. For additional duties during direct fire, see Section VII. Except in rapidly moving situations ammunition will be prepared on the ground as described in paragraph **50**.

 a. **Chief of section.** Supervises action of all members of the section.

 b. **Driver.** Remains at his post.

 c. **Gunner.** Lays the piece for direction.

 d. **Cannoneer No. 1.** Lays the piece for range, opens and closes the breech, and fires the piece.

 e. **Cannoneer No. 2.** Loads the piece.

 f. **Cannoneer No. 3.** Prepares ammunition for firing.

 g. **Cannoneer No. 4.** Assists No. 3 in preparing ammunition for firing.

20. CHIEF OF SECTION, LIST OF DUTIES. a. Relays all fire commands from the battery executive.

 b. Sees that all commands are executed rapidly and accurately.

 c. Measures the minimum quadrant elevation.

 d. Indicates to gunner the aiming point or referring point.

e. Lays for elevation, assisted by No. 1 when the gunner's quadrant is used.

f. Measures the range or elevation.

g. Indicates when the piece is ready to fire.

h. Signals or commands FIRE.

i. Reports errors and other unusual incidents of fire to the battery executive.

j. Conducts prearranged firing schedules.

k. Records basic data.

l. Frequently notes functioning of materiel.

m. Checks all rounds which have been prepared for firing but not fired, before they are replaced in containers.

n. Assigns duties under unusual conditions.

o. Controls movement of motor carriage to new firing positions.

21. CHIEF OF SECTION, DETAILED DESCRIPTION OF DUTIES. a. Relays all fire commands from the battery executive. The chief of section relays all fire commands received over intrabattery communication system. He is prepared to repeat any part of the command to any cannoneer who has failed to hear it.

b. Sees that all commands are executed rapidly and accurately. The chief of section frequently checks the laying of the piece and sees to it that all members of the section are working accurately and as fast as possible.

c. Measure the minimum quadrant elevation. The command is MEASURE THE MINIMUM QUADRANT ELEVATION. The chief of section, sighting along the lowest element of the bore, orders No. 1 to elevate the piece until the line of sight just clears the crest. The chief of section now sets the micrometer on the gunner's quadrant at zero and places the proper reference surface of the quadrant on the leveling plates of the breech ring parallel to the bore, with the "Line of Fire" arrow pointing toward the muzzle of the howitzer. He then disengages the plunger from the notches in the frame, raises the quadrant arm, and then lowers it slowly until the level bubble passes through the center of the level. He engages the plunger in the notches and turns the micrometer knob until the level bubble is accurately centered. The chief of section then reads the red or black figures according to the engraved instructions below the micrometer. After reading the angle the chief of section reports it to the battery executive, MINIMUM QUADRANT ELEVATION, NUMBER (SO AND SO), (SO MUCH).

d. Indicates to gunner the aiming point or referring point. Whenever an aiming point or referring point has been designated by the battery executive, the chief of section makes sure that he has properly identified the point in question and then calls off AIMING POINT IDENTIFIED. He then indicates it to the gunner. If there is any possibility of misunderstanding, the chief of section turns the panoramic telescope until the center horizontal and vertical cross hairs of the reticle are on the designated point. With the aid of the sight, the gunner is then able to identify the point and announce to the chief of section, AIMING POINT IDENTIFIED.

e. Lays for elevation, assisted by No. 1, when the gunner's quadrant is used. The command QUADRANT (SO MUCH) in the fire command indicates that the gunner's quadrant is to be used. The chief of section sets the announced elevation on the gunner's quadrant and when the piece is loaded and the breechblock closed, he places the quadrant on the leveling plates of the breech ring. He stands squarely opposite the side of the quadrant and holds it firmly on the leveling plates parallel to the axis of the bore. No. 1 elevates the piece until the quadrant bubble is centered, making sure that the last movement is in the direction in which it is most difficult to turn the elevating handwheel. The chief of section warns No. 1 when the bubble is approaching the center, in order that the proper leveling may be performed accurately.

f. **To measure elevation (range).** At the command MEASURE THE ELEVATION (RANGE), the piece having been laid, the chief of section causes No. 1 to set site 300 and, with the range drum knob, to center the longitudinal level bubble of the range quadrant. The chief of section then reads the elevation (range) set on the elevation scale (range drum) and announces the elevation (range) thus set; for example, ELEVATION (RANGE) NUMBER (SO AND SO), (SO MUCH).

g. **Indicates when the piece is ready to fire.** When the battery executive can see arm signals given by the chief of section, the chief of section will extend his right arm vertically as a signal that the piece is ready to fire. He gives this signal as soon as the gunner calls READY. When arm signals cannot be seen, the chief of section reports orally to the battery executive, NUMBER (SO AND SO), READY.

h. **Signals or commands FIRE.** The chief of section gives the signal to fire by dropping his right arm sharply to his side. When arm signals cannot be seen he orally commands NUMBER (SO AND SO), FIRE. The chief of section will not give the signal to fire until all cannoneers are in their proper places.

i. **Reports errors and other unusual incidents of fire to the battery executive.** If for any reason the piece cannot be fired, the chief of section will promptly report that fact to the battery executive, giving the reason NUMBER (SO AND SO), OUT (REASON). Whenever it is discovered that the piece has been fired with an error in laying, the chief of section will report at once the amount of error and whether corrected or not.

For example: NUMBER (SO AND SO) FIRED 40 MILS RIGHT, ERROR HAS (HAS NOT) BEEN CORRECTED. The gunner also reports any other unusual incidents that affect the service of the piece.

j. **Conducts prearranged firing schedules.** Whenever the execution of prearranged fire schedules is ordered, the chief of section conducts the fire of the piece in strict conformity to the schedule prescribed.

k. **Records basic data.** The chief of section records in a notebook such information as minimum elevations, base deflections including aiming points used, prearranged firing schedules when prepared schedules are not furnished, safety limits in elevation and deflection, number of rounds fired with the hour and date, and calibration corrections when appropriate.

l. **Notes frequently the functioning of materiel.** The chief of section notes the functioning of all materiel during firing. He occasionally checks the recoil oil index to see that the recoil mechanism contains the proper amount of oil and at all times carefully observes the functioning of the recoil system. He promptly reports to the battery executive any materiel that is not functioning properly.

m. **Checks all rounds which have been prepared for firing but not fired, before they are replaced in containers.** The chief of section personally checks all rounds not fired which have been prepared for firing to see that all seven increments are present in proper condition, assembled in proper numerical order and are of the proper lot number before they are placed in their containers. He sees that time fuzes are recut to "SAFE", and that safety pins and packing stops are replaced before the rounds are returned to the containers. He also checks to see that the lot number on the ammunition corresponds to the lot number on the container. He obtains a slip of paper on which an officer of the battery has certified, with his initials, that all required precautions as to checking increments have been taken and gives it to No. 3 for inclusion under the seal when unused ammunition is replaced in containers and sealed.

n. **Assigns duties under unusual conditions.** Whenever the personnel of the section serving the piece is temporarily reduced or other conditions require it, the chief of section assigns duties as will best facilitate the service of the piece. For example, the chief of section may direct the driver, No. 2 and No. 3 to assist with the carrying and uncrating of ammunition, or No. 2 may act as the driver for a large deflection shift when the driver is handling ammunition.

o. **Controls movement of the motor carriage to new firing positions.** When it is necessary to move the motor carriage to a new firing position, the chief of section instructs the driver to start the engine. He then controls the displacement of the motor carriage to the new position by touch signals.

22. DRIVER, LIST OF DUTIES. a. Remains at his post.

b. Moves motor carriage as directed by the chief of section or gunner.

23. DRIVER, DETAILED DESCRIPTION OF DUTIES. a. **Remains at his post.** At all times during firing, the driver remains in the driver's seat. He keeps the engine running until the piece is laid.

b. **Moves motor carriage as directed by the chief of section or gunner.** The driver starts the engine on command of the chief of section and moves the motor carriage as directed by touch signals from the chief of section or gunner. The driver must see to it that the mechanical condition of the engine is such that it will start quickly and easily at all times. When in action, the driver will warm up the engine from time to time, to keep it in condition for immediate starting.

24. GUNNER, LIST OF DUTIES. a. Sets or changes deflection.
  b. Applies deflection difference.
  c. Lays the piece for direction.
  d. Calls READY.
  e. Refers the piece.
  f. Records the base deflection.
  g. Measures the deflection.

25. GUNNER, DETAILED DESCRIPTION OF DUTIES. a. Sets or changes deflection. (1) **Set deflection.** When the piece was prepared for action, the elevation indexes, azimuth scale and azimuth micrometer on the panoramic telescope were set at zero. At the command, for example, DEFLECTION 1885, the gunner first sets the micrometer index opposite the zero or opposite the proper deflection correction of the fixed deflection scale if it is not already so set. He then pushes the throwout lever and turns the rotating head until the hundreds graduation (18 in this case) is opposite the azimuth scale index. He then releases the throwout lever and, grasping the azimuth worm knob with his left hand, turns the knob until the graduation 85 on the azimuth micrometer is opposite the index. Care must be taken that the direction in which the azimuth knob is turned results in the proper reading and not one that is 100 mils in error. The line of sight will then make a horizontal angle of 1885 mils with the axis of the bore.

(2) **Change deflection.** At the command RIGHT (SO MUCH), the gunner deducts the announced figure from the deflection he previously set on the azimuth scale and micrometer by decreasing the setting on the micrometer. At the command LEFT (SO MUCH), the gun-

ner adds the announced figure to the deflection he previously set on the azimuth scale and micrometer, increasing the setting on the micrometer index. The gunner makes these changes by turning the azimuth micrometer knob with his left hand.

b. Applies deflection difference. (1) If the command is, for example, ON NUMBER ONE, OPEN FIRE, the gunner on No. 1 makes no change, the gunner on No. 2 turns the top of the azimuth worm knob away from him and sets off 5 mils once, the gunner on No. 3 turns the top of the azimuth worm knob in a similar manner, except he sets off 5 mils twice, or a total of 10 mils, the gunner on No. 4 also turns the top of the azimuth worm knob in a similar manner, except that he sets off 5 mils three times, or a total of 15 mils.

(2) Should the command be, for example, ON NUMBER THREE, CLOSE ONE ZERO, the gunner on No. 1 turns the top of the azimuth worm knob away from him and sets off 10 mils twice, or a total of 20 mils, the gunner on No. 2 turns the top of the azimuth worm knob in a similar manner except that he sets off 10 mils once, the gunner on No. 3 makes no change, the gunner on No. 4 turns the top of the azimuth worm knob toward him and sets off 10 mils once.

(3) When a deflection change and a deflection difference are announced at the same time, for example, RIGHT, THREE ZERO, ON NUMBER ONE, CLOSE FIVE, both of which affect the gunner's piece, he should first set off the deflection change and then apply the deflection difference.

(4) In the methods described above, the gunner can reset the micrometer index opposite the most convenient 10-mil graduation after each deflection setting is made. When this is done he starts each change in the deflection setting with the micrometer index set at an even ten. This facilitates setting off the tens and units on the

azimuth micrometer, the micrometer index is turned to the zero or to the proper deflection on the deflection scale, and the setting opposite the arrow on the micrometer index is read.

(5) Due to the variety of formations of the armored artillery, it often will be necessary to order deflection shifts for each individual piece. The command will be given by the battery executive and will indicate the number of the piece and the deflection change to be made. For example, NUMBER THREE, RIGHT FIVE. The change will be made in the same manner as making any deflection change.

c. Lays the piece for direction. (1) With the deflection set on the panoramic telescope, the gunner traverses the piece until the center vertical cross hair of the panoramic telescope is approximately on the aiming point. He then centers the bubbles of the telescope mount. The gunner now brings the center vertical cross hair of the panoramic telescope into exact alignment with the aiming point and then rechecks the bubbles. To take up lost motion and insure accuracy, the final movement of the traversing handwheel should be such that the center vertical cross hair of the panoramic telescope is always brought onto exactly the same part of the aiming point from the left. The gunner should habitually follow the above procedure in laying the piece for direction.

(2) If the amount of movement necessary is greater than can be obtained by traversing the piece, the gunner will center traverse the howitzer and indicate by touch signals to the driver the movement required. Any movement of the carriage will involve a displacement from the line of aiming posts, if in use, and the gunner will continue to control the movement of the carriage until it is returned to its original position with the sight approximately on the line of aiming posts. Until cor-

rection for alignment is ordered he will traverse the piece until the far post appears on the horizontal cross hair of the reticle half-way between the vertical cross hair and the near aiming post.

d. Calls READY. When the piece is laid for direction, and No. 1 has called SET, the gunner moves his head clear of the eyeshield, checks the cross and longitudinal levels on the telescope mount, checks the setting of the azimuth scale and micrometer, and indicates to the chief of section that the piece is ready to fire by announcing READY.

e. Refers the piece. When the piece is laid for direction, to refer the piece the command is, AIMING POINT (SO AND SO), REFER. Without disturbing the laying of the piece, the gunner brings the center vertical cross hair of the panoramic telescope on to the new aiming point. He then reads and announces the deflection and notes it and the referring point on a convenient part of the motor carriage.

f. Records base deflection. At the command RECORD BASE DEFLECTION, the gunner reads the deflection then set on the panoramic telescope and announces it to the recorder as BASE DEFLECTION, NUMBER (SO AND SO), (SO MUCH).

g. Measures the deflection. The command is AIMING POINT (SO AND SO), MEASURE THE DEFLECTION. The piece being laid for direction, without disturbing the laying of the piece the gunner turns the panoramic telescope until the center vertical cross hair of the reticle is on the designated aiming point. He centers the bubbles on the telescope mount, rechecks line of sighting of the panoramic telescope, and then reads the deflection set on the azimuth scale and micrometer. He announces the reading as DEFLECTION NUMBER (SO AND SO), (SO MUCH).

26. CANNONEER NO. 1, LIST OF DUTIES. a. Sets angle of site.
   b. Sets range.
   c. Sets elevation.
   d. Lays the piece for range.
   e. Opens and closes the breech.
   f. Calls SET.
   g. Fires the piece.
   h. Mans the machine gun.

27. CANNONEER NO. 1, DETAILED DESCRIPTION OF DUTIES. a. Sets angle of site. The angle of site is set on the angle of site scale and micrometer of the range quadrant. No. 1 turns the angle of site worm knob until the announced figure is approximately indicated on the angle of site scale. He then makes a small final adjustment of the micrometer to bring the last two figures of the angle of site opposite the micrometer index. For example, the command is SI 275. No. 1 turns the angle of site worm knob until the scale index is one quarter the distance from the 3 to the 2 graduation. He then adjusts the micrometer to read 75. The last motion is made in the direction of increasing site.

   b. Sets range. Range is set on the range drum of the range quadrant. To set range, No. 1 first installs the range drum corresponding to the announced charge. He grasps the range drum knob in his right hand and turns it until the announced range is opposite the index, making sure that the last movement is in the direction of increasing range.

   c. Sets elevation. No. 1 turns the range drum knob until the announced elevation is approximately indicated on the elevation scale. By a small movement of the micrometer he then brings the last two figures of the elevation opposite the elevation micrometer index. He makes sure that the last movement is in the direction of increasing elevation. For example, the command is ELEVATION 464. No. 1 sets the elevation scale index between the four and five graduation. He then adjusts the micrometer to read 64. *Caution:* A large final adjustment of the micrometer indicates either an improperly adjusted elevation scale or an incorrect initial setting. A check should be made to see that a 100 mil error does not exist.

d. Lays the piece for range. After setting the range on the range drum or setting the elevation on the elevation scale and micrometer, No. 1 lays the piece for range. He first centers the cross level bubble by turning the cross leveling worm knob. Next, he elevates or depresses the piece by turning the elevating handwheel until the bubble in the longitudinal level is approximately centered. No. 1 now rechecks the cross level and centers the bubble again, if necessary. He now accurately centers the longitudinal level bubble, making sure that the final movement of the tube is in the direction in which it is most difficult to turn the elevating handwheel.

e. Opens and closes the breech. (1) *To open.* No. 1 grasps the breech operating lever handle in his left hand, pushes down on the handle to release the catch, and draws it toward him and to the rear, opening the breech.

(2) *To close.* No. 1 grasps the operating lever handle with his left hand, pushes handle forward and away from him until the breech is closed and the latch is engaged.

f. Calls SET. No. 1 calls SET when the piece has been loaded, the breech closed, and the piece laid for elevation or range.

g. Fires the piece. At the chief of section's signal or command NUMBER (SO AND SO), FIRE, No. 1 grasps the handle of the lanyard with his left hand and pulls it away from the piece as far as possible. In case of misfire, the instructions contained in paragraph **60** will be followed.

h. Mans the machine gun. The machine gun will be operated by No. 1 when so directed by the chief of section. When the motor carriage is traveling, No. 1 will remain on the alert, prepared to operate the machine gun instantly.

28. CANNONEER NO. 2, LIST OF DUTIES. a. Loads the piece.
   b. Calls out number of round.
   c. Inspects chamber and bore between each round.
   d. Sets out aiming posts when ordered.
   e. Disposes of used cartridge cases.

29. CANNONEER NO. 2, DETAILED DESCRIPTION OF DUTIES. a. Loads the piece. No. 2 picks up the round from the left side of the rear deck of the motor carriage, or receives it from No. 4. He grasps it with his right hand at the base of the cartridge case and his left hand in front of the rotating band. He then faces the breech and inserts the round into the chamber and pushes it home with his right hand. He must use care especially at higher elevations, to avoid injuring his hand. When it is necessary for No. 2 to insert his hand into the breech recess to push a round home, he should first close his fist. No. 2 will be particularly careful to avoid striking the fuze against any portion of the materiel. A round to be loaded will be held well out of the path of the recoil of the howitzer until the howitzer returns to battery. (See AR 750-10.)

b. Calls out number of round. When the method of fire is other than one round, No. 2 calls out the range and the number of the round as he loads the piece;

and as he loads the last round, adds LAST ROUND. He should not speak louder than necessary for insuring his being heard by members of his own howitzer section.

c. **Inspects chamber and bore between each round.** No. 2 will inspect the bore and chamber between each round to make certain that no foreign matter has been left from previous rounds.

d. **Sets out aiming posts when ordered.** When so directed by the chief of section, No. 2 dismounts from the motor carriage and takes the aiming posts from the rear deck of the motor carriage. He sets up the aiming posts under the guidance of the gunner. (See paraparagraph 55.)

e. **Disposes of used cartridge cases.** During lulls in firing, No. 2 will throw used cartridge cases over the right side of the motor carriage.

30. **CANNONEER NO. 3, LIST OF DUTIES. a.** Places packaged ammunition on rear deck.

b. Removes round from container.

c. Cleans and inspects projectile.

d. Sets fuze setter.

e. Cuts or sets fuzes.

f. Resets fuzes and replaces in containers rounds prepared for firing but not fired.

31. **CANNONEER No. 3, DETAILED DESCRIPTION OF DUTIES. a. Place packaged ammunition on rear deck.** As soon as No. 3 knows what type of ammunition is to be fired, he takes a packaged round from the ammunition rack and places it, projectile end up, on the rear deck of the motor carriage.

b. **Removes round from container.** No. 3 strips the tape from the projectile end of the ammunition container as shown in figure 8, and throws the end and tape away. He picks up the container and dumps the projectile into No. 4's waiting hands, as shown in figure 9. No. 3 then reverses the container and strips the tape from the other end, as shown in figure 10, throwing the tape and end away. He now dumps the cartridge case into No. 4's hands, as shown in figure 11, and disposes of the container.

c. Cleans and inspects projectile. No. 3 wipes off the projectile with a clean rag or waste and carefully inspects the rotating band for burrs. If it is burred, No. 3 sets it aside until he has an opportunity to file down the burrs.

d. Sets fuze setter. When time fuzes are to be used, No. 3 sets the corrector and time announced in the fire command on the fuze setter so that it is ready for instant use. He also makes announced changes during the firing.

e. Cuts or sets fuzes. (1) *Time fuzes.* No. 3 removes the safety pin and cuts the fuze to time ordered, using the fuze setter as shown in figure 12.

(2) *Impact fuzes.* No. 3 sets the adjustable impact fuzes to "QUICK" (SQ) or "DELAY" (D), as ordered. Fuzes are originally furnished set "QUICK", but this setting must be verified in each case.

f. Resets fuzes and replaces in containers the rounds prepared for firing but not fired. Under the personal supervision of the chief of section, No. 3 resets time fuzes to "SAFE", replaces the safety pin, and replaces in the containers the projectiles and cartridge cases of all ammunition that has been prepared for firing but not fired. He is careful that the lot number on the ammunition corresponds to the lot number on the container and that the packing stop and cardboard cover are included. He includes the slip of paper (Officer's Certificate) obtained from the chief of section, under

Figure 8. No. 3. opening projectile end of ammunition container.

one of the seals in a visible position when resealing the containers.

32. CANNONEER No. 4, LIST OF DUTIES. a. Receives projectile and cartridge case from No. 3.

b. Cleans and inspects cartridge case.

c. Prepares charges.

d. Assembles rounds.

e. Places prepared rounds on left side of rear deck.

f. Replaces increments in cartridge cases of all rounds prepared for firing but not fired.

Figure 9. No. 4 receiving projectile from ammunition container held by No. 3.

Figure 10. No. 3 opening cartridge case end of ammunition container.

Figure 11. No. 4 receiving cartridge case from ammunition container held by No. 3.

33. CANNONEER NO. 4, DETAILED DESCRIPTION OF DUTIES. a. Receives projectile and cartridge case from No. 3. When No. 3 opens the projectile end of the ammunition container, he dumps the fuzed projectile into No. 4's hands, as shown in figure 9. No. 4 lays the projectile down in the trough pointing toward the right side of the motor carriage (No. 4's left as he faces rear deck). He then receives the cartridge case from the container which is still held by No. 3, as shown in figure 11.

b. Cleans and inspects cartridge case. No. 4 wipes off the cartridge case with a clean rag or waste. While doing this, he carefully inspects the cartridge case to make sure that it is not bent or damaged. If the cartridge case is in such condition that it cannot be used,

Figure 12. No. 3 cutting fuze while No. 4 fixes charge.

Figure 13. No. 4 assembling round.

Figure 14. No. 4 placing round on rear deck, projectile pointing to rear.

it is set aside until it can be straightened or disposed of.

c. **Prepares charges.** The fire command designates the charge to be fired. When No. 4 receives the cartridge case from No. 3, he withdraws the increments as shown in figure 12, removing those numbered higher than the charge designated, and disposes of them. He replaces the remaining increments in the cartridge case in their original numerical order.

d. **Assembles rounds.** When No. 4 has prepared the charge and No. 3 has set or cut the fuze, No. 4 assembles the projectile and cartridge case in the ammunition trough as shown in figure 13.

e. **Places prepared rounds on left side of rear deck.** When No. 4 assembles a round he places it on the left side (No. 4's right) of the rear deck of the motor carriage with the projectile pointed to the rear, ready for No. 2 to pick up and load into the breech. When No. 2 is ready to load the piece and no rounds are prepared, No. 4 passes the prepared rounds to No. 2. To do this, when No. 4 has assembled a round he places his right hand under the cartridge case and his left hand under the projectile, making sure that the projectile and cartridge case do not separate. He passes the round to No. 2 in such a manner that No. 2 is able to grasp the base of the cartridge case in his right hand.

f. Replaces increments in cartridge cases of all rounds prepared for firing but not fired. Under the personal supervision of the chief of section, No. 4 replaces in cartridge cases a full charge of increments for all ammunition prepared for firing but not fired. No. 4 carefully checks to see that all seven increments are present, in the proper condition, assembled in the proper numerical order, and that they are of the proper lot number.

Section VII

DIRECT FIRE

34. GENERAL. Delivery of direct fire demands a high degree of training in its special technique, since it requires the section to operate as an independent unit. This training is based on the technique employed in the normal mission of indirect fire. The high standards of speed and accuracy required in indirect fire are even more important during direct fire on a target which may, in turn, be firing on you. Three systems of distributing the duties of individuals in laying the piece for direct fire are employed with this weapon. All systems should be equally emphasized and for any given target the system selected should be that which will provide the most rapid means of obtaining effect.

35. PREPARATORY STEPS. In order to obtain the best results when using direct fire the following preparatory steps should be taken:

a. Permanently attach range conversion charts, similar to those shown in figures 15 and 16, to the inside of the fighting compartment of the motor carriage. Since Shell, HE, Charge VII, and HEAT are used with this weapon and the sight reticles of the elbow telescope M16 and the panoramic telescope M12A2, are graduated for Shell, HE, Charge V, range conversion charts are essential. Figure 15 shows the chart which should be installed in a convenient position for reference by the

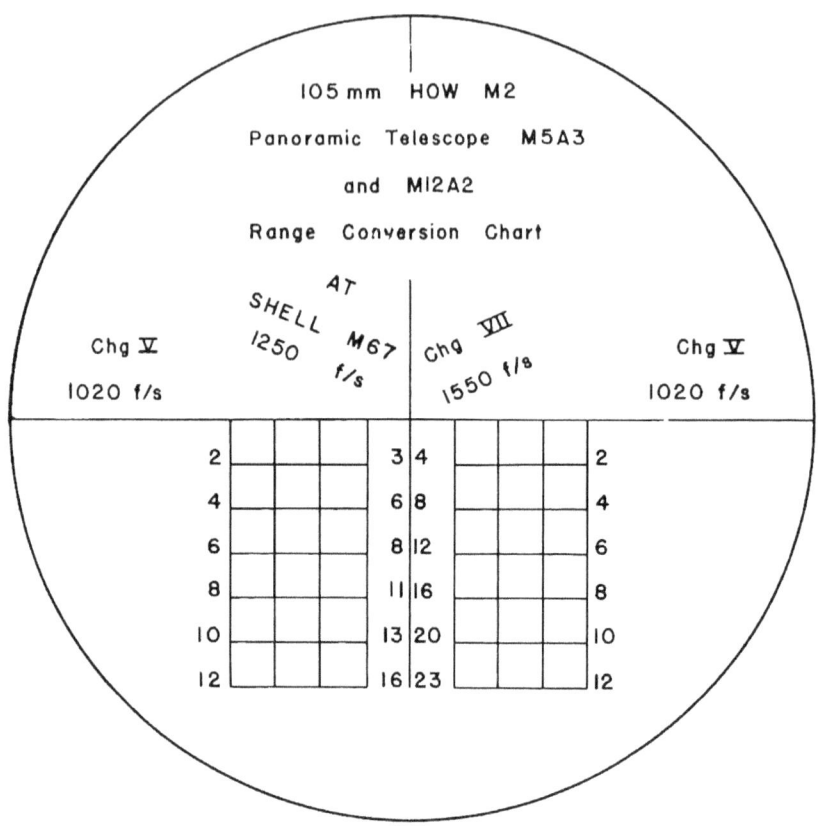

Figure 15. Range conversion chart, panoramic telescope M5A3.

gunner. Figure 16 shows the chart that will be used by No. 1. The elbow telescope M16A1C has reticle graduated for Shell, HE, Charge VII, and HEAT. This reticle is illustrated in figure 17.

b. Inspect the panoramic and elbow telescopes and mounts for looseness, and have them as tight as consistent with free operation.

c. Boresight both panoramic and elbow telescopes with trunnions as level as possible.

36. **BORESIGHTING.** a. Accurate fire of any type requires accurate adjustment of sighting and laying equipment. All personnel must be impressed with the importance of accurate boresighting at every opportunity. The tests and adjustments described in Section IX of this manual and in TM 9-731E will insure accurate sight adjustment for direct laying. The use of the sight extension bar on the panoramic telescope mount introduces sufficient whip, when the piece is fired, to throw the elevation indexes on the mount out of coincidence. The

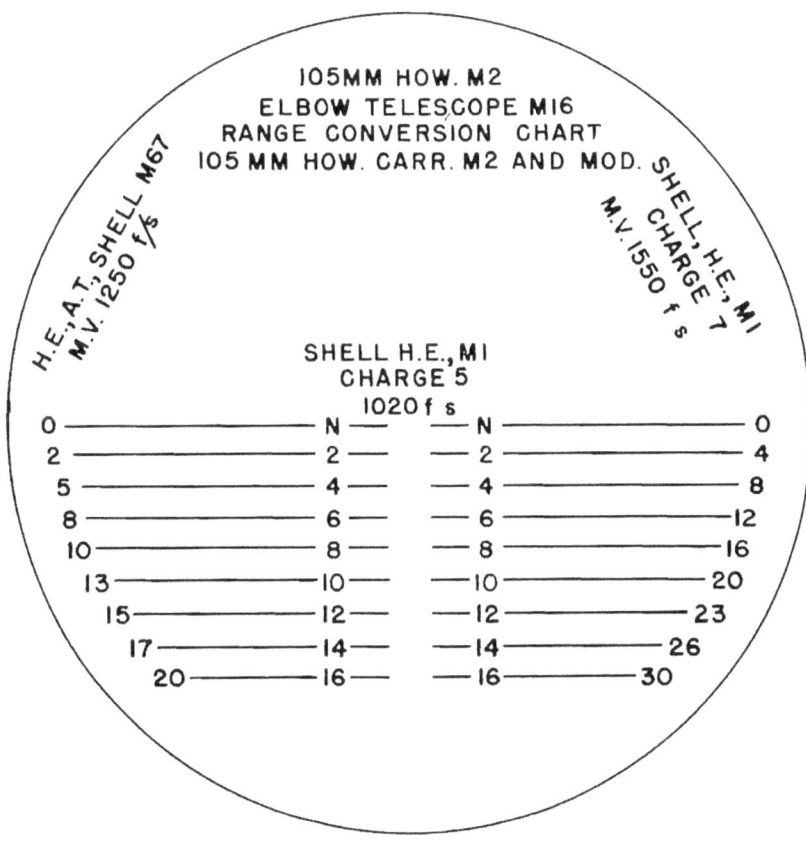

Figure 16. Range conversion chart, elbow telescope M16.

57

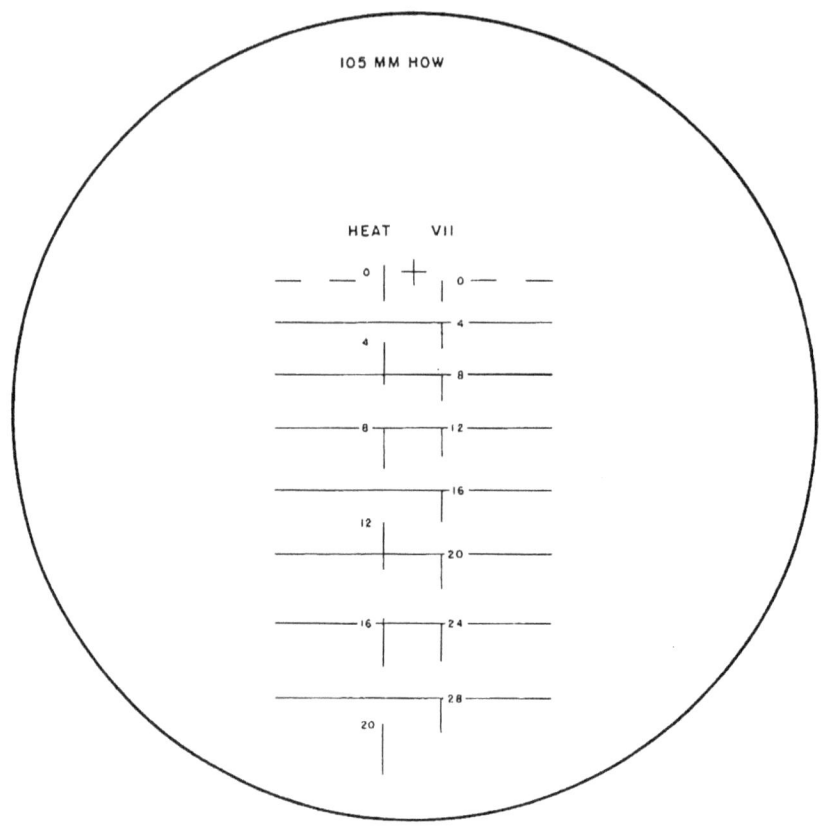

Figure 17. Reticle, elbow telescope M16A1C.

indexes themselves are sufficiently thick to prevent extreme accurate matching by eye.

b. When the one-man one-sight system of direct laying is employed, the plane through the center horizontal cross hair of the panoramic telescope must be parallel to a plane through the horizontal axis of the tube. In boresighting, this is accomplished by first leveling the tube with the gunner's quadrant or range quadrant, and then leveling the panoramic sight mount by centering the longitudinal level bubble. This makes the sight mount and tube parallel. Then panoramic telescope is adjusted by boresighting as indicated in paragraph **68**.

c. Prior to firing the first round, the tube and sight mount are made parallel in the above manner. For subsequent rounds the sight mount indexes must be checked and realigned after each round is fired in order to keep the sight mount parallel to the tube.

d. On those vehicles that are modified to provide a dowelled locking pin on the hub of the telescope mount, the mount can be locked in adjustment with the tube for direct laying by approximately aligning the indexes and rotating the elevating knob of the mount until the pin can be engaged in the hole and locked. This pin will keep the mount in adjustment during firing. It is installed so that when it is locked in place the mount and tube are in adjustment.

37. SYSTEMS OF DIRECT FIRE. a. **One-man one-sight system.** The gunner lays for both direction and range using the gridded reticle of the panoramic telescope.

b. **Modified one-man one-sight system.** This system is similar to the one-man one-sight system except that No. 1 lays for range with the range quadrant while the gunner lays for direction with the gridded reticle of the panoramic telescope. After laying the piece initially, subsequent changes in elevation are made by using the graduations on the elevating handwheel.

c. **Two-man two-sight system.** The gunner lays for direction using the gridded reticle of the panoramic telescope while No. 1 lays for range using the range lines of the elbow telescope.

d. The one-man one-sight system offers this advantage: since one man, the gunner, lays the piece for both direction and range, there is small probability of the piece being laid on the wrong target. The two-man two-sight system offers the advantage of a faster rate of fire; that is, more aimed shots per minute can be delivered than with the one-man one-sight system. Training should include both systems. The relative advantages and disadvantages of the two systems should be considered in electing which to use in any given situation.

38. THE ONE-MAN ONE-SIGHT SYSTEM OF DIRECT FIRE. a. Duties of the chief of section. (1) Conducts the fire of the piece.

(2) Prepares a range card for his sector and corrects card by data obtained from firing.

(3) Provides for the clearance of all immediate obstructions within his sector.

(4) Selects or identifies the target.

(5) Estimates the range to the target.

(6) Determines the lead in mils for moving targets.

(7) Gives the initial fire command.

(8) Senses each round and gives commands for changes in range and lead or deflection.

(9) Checks and realigns indexes of panoramic telescope mount after each round is fired.

b. Duties of the gunner. (1) Levels the panoramic telescope by centering the longitudinal level bubble of the telescope mount.

(2) Sets the elevation indexes on the panoramic telescope at zero.

(3) Sets the azimuth scale and micrometer of the panoramic telescope at zero.

(4) Lays the piece approximately for direction.

(5) Cross levels the panoramic telescope mount and keeps the cross level bubble centered during firing.

(6) Lays the center vertical cross hair accurately on the center of the target.

(7) Lays the piece for range by laying the proper range line accurately on the center of the target.

(8) If a moving target, tracks with the announced lead.

(9) Calls FIRE after No. 1 has called SET, and the piece is laid.

(10) Continues tracking the target while howitzer is being fired.

(11) Applies corrections for range and lead or deflection as announced by chief of section.

c. Duties of Cannoneer No. 1. (1) Levels the howitzer using the longitudinal level bubble of the range quadrant.

(2) Opens and closes the breech.

(3) Calls SET when the piece is loaded.

(4) Fires the piece.

Figure 18. The one-man one-sight system of direct fire.

d. Duties of Cannoneer No. 2. (1) Takes round from rear deck, or receives it from No. 4.

(2) Loads the piece.

e. Duties of Cannoneer No. 3. Prepares ammunition for firing.

f. Duties of Cannoneer No. 4. (1) Prepares ammunition for firing.

(2) Places prepared rounds on rear deck, or passes them to No. 2.

g. Duties of driver. Remains at his post.

39. THE MODIFIED ONE-MAN ONE-SIGHT SYSTEM. In the modified one-man one-sight system of direct fire the gunner lays the piece for direction with the panoramic telescope while No. 1 lays the piece for range using the range quadrant and the graduated elevating handwheel. The chief of section determines the elevation from the firing tables. If there is an angle of site, he adds it to the elevation shown in the firing table before announcing the elevation to No. 1. No. 1 sets the angle of site scale at 300, sets the elevation scale and micrometer at the announced elevation, centers the cross level bubble and then lays the piece for range by turning the elevating handwheel until the longitudinal level bubble is centered. Subsequent changes in elevation are made with the graduated elevating handwheel. If time permits, the chief of section measures the angle of site to several points in his sector for later use as reference points.

40. THE TWO-MAN TWO-SIGHT SYSTEM OF DIRECT FIRE. a. Duties of the chief of section. (1) Conducts the fire of the piece.

(2) Prepares a range card for his sector and corrects card by data obtained from firing.

(3) Provides for the clearance of all immediate obstructions within his sector.

(4) Selects or identifies the target.

(5) Estimates the range to the target.

(6) Determines the lead in mils for moving targets.

(7) Gives the initial fire command.

(8) Senses each round and gives commands for changes in range and lead or deflection.

b. Duties of the gunner. (1) Levels the cross level bubble on the panoramic telescope mount.

(2) Sets the elevation indexes on the panoramic telescope at zero.

(3) Sets the azimuth scale and micrometer of the panoramic telescope at zero.

(4) Lays the piece for direction.

(5) Tracks moving target with center vertical cross hair in center of target, then takes the announced lead.

(6) Commands FIRE after No. 1 has called SET.

(7) Continues tracking target while howitzer is being fired.

(8) Applies corrections for lead or deflection, as announced by chief of section.

c. Duties of Cannoneer No. 1. (1) Lays the piece for range with the elbow telescope by laying proper range line accurately on the center of the target.

(2) Tracks moving targets by keeping the proper range line on the center of the target at all times.

(3) Calls SET when the piece is loaded and laid for range.

(4) Applies range changes announced by chief of section.

d. Duties of Cannoneer No. 2. Loads the piece.

e. Duties of Cannoneer No. 3. (1) Opens and closes the breech.

(2) Fires the piece on command from gunner.

Figure 19. The two-man two-sight system of direct fire.

f. Duties of Cannoneer No. 4. (1) Prepares ammunition for firing.

(2) Places prepared rounds on rear deck of motor carriage.

g. Duties of the driver. Remains at his post.

41. FIRE COMMANDS. Fire commands for direct fire are given by the gun commander who normally is the chief of section. The battery executive designates the piece or pieces to fire, and the target, followed by the command FIRE AT WILL. When the chief of section receives this preliminary command, he gives a fire command to his section containing, in sequence, the following:

Alert and system of fire,
Target,
Ammunition,
Direction of target,
Lead,
Range of elevation.

42. ALERT AND SYSTEM OF FIRE. The command to alert the section is (NUMBER) SECTION. The system of fire will be the one-man one-sight system unless others designated by the chief of section or the battery executive. If the two-man two-sight system or modified one-man one-sight system is to be used, the chief of section adds to the alert command TWO-SIGHT or MODIFIED ONE-SIGHT. This allows the section to move to its proper position so that the crew is ready for the initial fire command. If the target is outside the maximum traverse of the howitzer, the alert is followed by the necessary signals to the driver to get the howitzer tube pointed in the approximate direction of the target.

43. TARGET. This section of the command is to identify the target. The identification must be unmistakable and should employ the minimum number of words. For example, TANK, MOVING TANK, MACHINE GUN, ANTITANK GUN or ENEMY INFANTRY (any enemy dismounted personnel).

44. AMMUNITION. a. Shell, HE. The command is SHELL, HE, FUZE DELAY or SHELL, HE, FUZE QUICK. Since Charge VII is always used during direct fire, no command for charge is necessary. Fuze delay should be used most frequently, as more penetration will be obtained on armored targets or fortifications. Against personnel, the ricochet produced by the 0.15

second delay gives an effective burst. If less than 50 per cent of the rounds fired against personnel are ricochet bursts, change to fuze quick.

   b. **Shell, HEAT.** The command is SHELL, ANTITANK. Since the charge and fuze are fixed, no other command is necessary. Shell, HEAT, is a lower velocity shell than HE, Charge VII, therefore its trajectory has a greater curve. It rarely detonates except when it hits armor or similar hard surface. This makes it harder to obtain hits against tanks with it, due to the difficulty of sensing the point of impact of a round which does not hit an armored target. It will pierce 5.5 inches of armor. The amount of HEAT carried is limited. In view of these facts, HEAT should be used only for close defense and should not be used at ranges in excess of 700 yards.

45. DIRECTION OF TARGET. The command is TRAVERSE RIGHT (LEFT), STEADY---ON. The chief of section while giving these commands, looks down the tube or through the open sight on the panoramic telescope. If the gunner has difficulty in locating the target, the chief of section may look through the panoramic telescope himself and put the vertical cross hair on the target.

46. LEAD. The initial command for lead is LEAD (SO MANY MILS). During adjustment changes in lead are announced as RIGHT (LEFT) (SO MANY MILS). The table (figure 20) shows the leads which should be used. It must be impressed upon all gunners and chiefs of section that a round which misses a tank for deflection, but is correct for range, will burst beyond the tank. When a tank approaches head-on, a hit should be obtained on the first or second round, and must be obtained on the third round.

SPEED	LEAD			
	TARGET TRAVELLING PERPENDICULAR TO LINE OF FIRE.		TARGET TRAVELLING 45° TO LINE OF FIRE.	
	HE-AT	CHARGE VII	HE-AT	CHARGE VII
SLOW (0-10 MPH)	5	5	5	5
MEDIUM (10-20 MPH)	20	15	15	10
FAST (OVER 20 MPH)	30	25	20	20

Figure 20. Lead table.

47. RANGE OR ELEVATION. The initial command for range is (SO MANY) YARDS, or elevation (SO MANY) MILS. During adjustment of fire, range or elevation is corrected by the command ADD (DROP) (SO MANY YARDS) or ADD (DROP) (SO MANY MILS).

48. SUBSEQUENT FIRE COMMANDS. a. Sensing and range changes. (1) The chief of section senses each round and announces the necessary changes in range and lead. When shell HE with charge VII is fired, the following trajectory characteristics will govern the manner of conducting fire:

(a) *Ranges from 0 to 500 yards.* Within these range limits the trajectory will be too flat to permit an 8-foot tank to pass under it. The upper range of 500 yards is the ideal at which to open fire on an approaching tank, since rapid fire can then be conducted without misses if deflection is correct.

(b) *Ranges from 500 to 1100 yards.* These range limits include the zone in which the trajectory is sufficiently flat to permit direct estimation of errors without actual bracketing of the target. Assuming zero vertical dispersion, if a hit is obtained on the bottom of an 8-foot tank at the upper limit (1100 yards), a 100-yard range change (to 1200 yards) will result in a round which will just brush the top of the tank. During adjustment within this zone, range changes should seldom exceed 100 yards and frequently ranges of 50 yards will be sufficient. The upper limit is the greatest range at which fire should be opened unless tactical conditions require otherwise. The second shot (or certainly the third) should be a hit.

(c) *Ranges from 1100 to 1800 yards.* The ranges from 1100 to 1800 yards represent a zone in which hits are reasonably possible. Ordinarily, bracket methods are used to obtain an adjustment for range in this zone. Fire should not be opened at these ranges unless surprise is of no consideration. Dispersion is a considerable factor in firing at targets within this zone.

(d) *Ranges over 1800 yards.* At ranges above 1800 yards, direct fire is not advisable against ordinary targets. However, it would be pointless to withhold fire until the target came within a range of 1800 yards if the gun position had been disclosed. Dispersion is the controlling factor. Ranges must be known accurately or determined by bracketing. At ranges above 1800 yards, fire becomes plunging and moving targets can avoid it easily.

(e) At short ranges, a round beyond the target is not necessarily incorrect for range, if deflection is wrong, since the point of aim is above the ground level.

(f) On targets smaller than an 8-foot tank, bracketing becomes necessary at ranges greater than 500 yards. For example, in firing at a hull-down tank with a vertical profile of 3 feet, it is necessary to bracket at a range of 900 yards, unless the range to the target is known accurately.

(2) The vertical displacement obtained with the 105-mm howitzer M7 for a 100-yard range change increases rapidly as the range to the target increases. In figure 21, the target is at a range of 500 yards. A prop-

Figure 21. Vertical changes in feet for 100-yard range changes at 500-yard target range. 105-mm howitzer M7, shell HE, M1, charge VII.

erly aimed round fired at a range of 500 yards will hit at A, the center of mass. Round B fired at 600 yards will strike the target 3 feet above A. Decreasing the range to 400 yards will lower the point of impact 3 feet, and round C will strike 1 foot above the base of the tank.

b. **Examples.** The following are some examples of typical fire commands.

(1) *Initial commands*
(a) *Stationary target (one-man one-sight system)*
    THIRD SECTION
    MACHINE GUN
    SHELL, HE, FUZE, DELAY
    TRAVERSE RIGHT
    STEADY --- ON
    800
(b) *Moving target (two-man two-sight system)*
    FIFTH SECTION
    TWO-SIGHT
    MOVING TANK
    SHELL, ANTITANK
    TRAVERSE LEFT
    STEADY --- ON
    LEAD 15
    400

(2) *Subsequent command*
(a) *Stationary target:*
1. SHORT
    RIGHT 4
    ADD 8 (using mils on graduated elevating handwheel)

2. DOUBTFUL
   LEFT 12
   REPEAT RANGE

(b) *Moving target:*
1. OVER
   DROP 200 (using sight reticle in yards)
2. Short
   LEFT 3 (change in lead)
   ADD 100

49. DISPERSION. Vertical dispersion is not a critical factor in adjustment of direct fire until the dispersion pattern approaches the size of the target. With Charge VII, Shell HE has a total dispersion pattern of 7½ feet at 1200 yards. A small error in range or in use of the sights would throw some rounds off the target in this case. At ranges where the dispersion pattern is as large as or larger than the height of the target, a range which has previously proved correct is not changed on the evidence of one round.

50. NIGHT FIRING. Using night lighting devices for sighting equipment, it is possible to use normal methods of direct laying when the target is sufficiently illuminated by moonlight or flares. The blinding effect of the muzzle blast can be overcome by closing the eyes as the piece is fired. If the flash of an enemy weapon is observed, it may be possible to lay the piece for direction by sighting on the flash. Range estimation will be difficult but if the section was carefully studied during daylight, and a range data card prepared, a satisfactory initial range may be determined by observing the flash in relation to terrain features. If time permits, the short base method range estimation, using the aiming circle, produces very accurate results.

Section VIII

ADDITIONAL INFORMATION ON THE SERVICE OF THE PIECE

51. PREPARING AMMUNITION ON THE GROUND. In situations where the howitzer is dug in, or where large quantities of ammunition will be fired from one position, it will often be necessary to prepare the ammunition on the ground. In this case No. 4 and the driver dismount to the left rear side of the motor carriage. They spread the ammunition tarpaulin on the ground and place the ammunition trough on it. Assisted by ammunition handlers, if available, No. 4 and the driver then remove ammunition from containers, clean the projectiles and cartridge cases, inspect the rotating bands, fix the charges and assemble the rounds. No. 4 then passes the rounds to No. 3 who remains in the motor carriage. No. 3 receives the rounds, sets or cuts the fuzes, and either passes the round to No. 2 or places it on the rear deck of the motor carriage, projectile pointing to the rear, where No. 2 can pick it up.

52. REPACKING AMMUNITION IN CONTAINERS.
a. Great care must be exercised in returning ammunition to containers to insure that it is completely serviceable. Before a round is replaced in its container, an officer of the battery will prepare a slip of paper on which he certifies, by his initials, that these checks have been made:
   (1) All seven increments present.

(2) Increments serviceable and undamaged.
(3) Increments in proper numerical order.
(4) Increments have same lot number.
(5) Lot number on ammunition same as on container.

b. The slip of paper bearing the officer's initials will be inserted under the sealing tape used to seal the container. It will be visible after sealing. Only rounds for which this certificate can properly be made will be repacked and sealed. If rounds must be repacked when the officer is not present, they will be so marked and held for later inspection to determine if the certified packing slip may be attached.

53. CARE OF AMMUNITION. Ammunition, especially the rotating bands and cartridge cases, must be protected from damage. It is sorted and stored by lots. It is kept in containers as long as practicable. Whether in or out of containers, it is protected from dirt and ground moisture by being placed on tarpaulins or raised off the ground.

54. ACCURACY IN LAYING. Sighting and laying instruments, fuze setters and elevating and traversing mechanisms will be so operated as to minimize the effects of lost motion. This requires that last motions in setting instruments and in laying be always in the directions prescribed. The gunner and No. 1 will invariably be required to verify the laying of the piece after the breech has been closed.

55. AIMING POSTS. When a suitable natural aiming point is not visible, the piece after it has been laid initially for direction is referred to the aiming posts as described in paragraph **24e.** Two aiming posts are used for each piece. Each post is equipped with a lamp for use in firing at night. One post is set up in a convenient location at least 100 yards from the piece. The other post is set up at the midpoint between the first post and the piece and is lined in by the gunner so that the center vertical cross hair of the panoramic telescope and the two aiming posts are all in line. Any lateral displacement of the piece during firing can then be detected easily and corrected as indicated in paragraph **56**. For night use, the lamps should be adjusted so that the far one will appear several feet higher than the near one; thus the two lamps will clearly establish a vertical line on which the center vertical cross hair of the panoramic telescope can be laid.

56. DISPLACEMENT CORRECTION. a. When a gunner sees that his aiming posts are out of line he notifies the chief of section (who notifies the executive) and uses the far post only for laying until a correction is authorized by the executive. The correction is made by the gunner, who—
  (1) Lays on the far post.
  (2) Refers to the near post.
  (3) Lays on the far post with the new reading.
  (4) Reports the new deflection.
  (5) Realigns the posts (as soon as practicable) by having the near post moved.

b. It is advisable to use this method of correcting displacement when making the final correction necessary to bring the panoramic telescope into alignment with the aiming posts, rather than attempting to make this final correction by continued movement of the motor carriage.

57. REPORTING ERRORS. All members of the howitzer section are trained to report to the chief of section errors in setting or laying discovered after the command **FIRE** has been given. The chief of section will immediately report errors to the executive, as prescribed in paragraph **21i**.

58. CEASE FIRING. The command CEASE FIRING normally is given to the howitzer section by the chief of section, but in emergencies anyone present may give the command. At this command, regardless of its source, firing will cease immediately. If the piece is loaded the chief of section will report that fact to the executive. Firing is resumed at the executive's announcement of the range or elevation.

59. SUSPEND FIRING. The command SUSPEND FIRING is given only when the battery is firing on a prearranged schedule and a temporary halt in the firing is desired. At this command firing is stopped, but settings continue to be altered in conformity with the schedule. If the piece is loaded, the chief of section will report that fact to the executive. Firing will be resumed at the executive's command RESUME FIRING.

60. TO UNLOAD THE PIECE. a. When the command UNLOAD is given, No. 1 opens the breech slowly. No. 2, standing at the breech, receives the ejected round or cartridge case.

b. Should the extractor fail to eject the complete round, the assembled staff and rammer (or staff and unloading device if available) is used. After seeing to it that the recess in the head of the rammer or device is free from obstructions and is designed to clear the type of fuze being used, No. 1 inserts the rammer or device into the bore until the head incloses the fuze and comes in contact with the projectile. He pushes and if necessary, taps the rammer staff lightly until the round is dislodged from its seat. In using the staff, the body must be kept clear of the front of the bore. He then pushes the projectile out of the breech. No. 2 receives it.

c. If the extractor has ejected the cartridge case but not the projectile, No. 1 fills the chamber with waste and closes the breechblock. He dislodges the projectile as described in **b**, above. No. 2 then opens the breech, removes the waste, and receives the projectile as No. 1 pushes it to the rear.

d. When practicable, the procedure prescribed in TM 9-1900 should be followed.

e. Where the use of the rammer-staff is necessary the piece will be unloaded under the supervision of an officer, unless in actual combat.

61. MISFIRES. In the event of a misfire, two attempts will be made to fire the piece before the breech is opened and the round removed. At the command UNLOAD, the procedure is the same as in paragraph **60**. If the extractor ejects the round, the round will be disposed of as prescribed in TM 9-1900. If the extractor ejects only the cartridge case (which will happen most frequently) the case will be immediately thrown clear of all personnel to prevent injury in case of a hangfire. Another cartridge case with the proper charge will be inserted in the breech, care being taken not to damage the case.

Section IX

CARE AND MAINTENANCE OF MATERIAL

62. GENERAL. a. This section covers the maintenance operations that may be performed by an armored field artillery battery in the field.

b. Complete instructions for battery maintenance are found in the Technical Manuals and Standard Nomenclature Lists referred to in paragraph 2. Operations not covered in those publications are performed by ordnance personnel.

c. In general, the battery is charged with preventive maintenance; that is, with routine cleaning and lubricating. Certain classes of minor repairs, adjustments, and replacement of parts may be made under the direction of the executive officer or the artillery mechanic. Parts which may be drawn by a battery for replacement purposes are those listed in the Organizational Spare Parts and Equipment Section of SNL C-21. Unless specific instructions are issued to the contrary all of these parts may be installed by the artillery mechanic. For routine care and maintenance, specific duties are assigned to individuals or howitzer sections and a strict accountability for the proper performance of such duties should be enforced.

d. The following operations may be performed within the battery:

(1) Draining and filling recoil cylinder.
(2) Dismounting the howitzer from the sleigh.

(3) Dismounting the sleigh from the cradle.

(4) Removal, disassembly and replacement of parts of the breechblock, breech operating mechanism, and firing mechanism.

(5) Removal of the equilibrator for cleaning and lubricating.

(6) Replacement of minor parts or assemblies listed in the Organizational Spare Parts section of SNL C-21.

63. INSPECTION. Regulations do not require that periodic inspections of ordnance material be made by field artillery troops. These inspections are performed by ordnance technical personnel. However, the battery executive should make a daily general inspection and a monthly detailed inspection of the materiel. The purpose of the daily inspection is to insure that the battery is properly performing its preventive maintenance functions. For this purpose, a general inspection of the motor carriage and weapon for appearance and a spot check of one or two parts are sufficient. At the monthly inspection all personnel of the firing battery should be present. The executive should make a thorough mechanical inspection of the motor carriage and he should inspect the ordnance tool sets, spare parts, and equipment for condition and completeness. The artillery mechanic should accompany the executive at the inspection of the weapons. All necessary repairs or adjustments which may be performed by the artillery mechanic should be accomplished; all other necessary repairs or adjustments should be reported to ordnance personnel.

64. CLEANING. Dirt and grit accumulated in traveling or from the blast of the piece in firing settle on the bearing surfaces, and in combination with the lubricant form a cutting compound. Primer fouling attracts moisture and hastens the formation of rust. During lulls in firing and immediately after firing, the piece must be thoroughly cleaned. At other times it should be cleaned at intervals, which may be daily, depending upon the use and condition. Dirt on nonbearing surfaces can usually be removed by water; lubricated or other greasy parts must be cleaned with dry-cleaning solvent applied with a rag. The procedure in cleaning the bore and breech mechanism is described in paragraph **67**. The cleaning materials issued by the Ordnance Department are listed in TM 9-850.

65. LUBRICATION. For lubrication instructions see the latest lubrication order for the equipment.

66. RECOIL MECHANISM. a. General. Battery maintenance of the recoil mechanism is limited to exterior cleaning and lubricating, draining and filling with recoil oil, removing the recoil cylinder rear head, and disconnecting the piston rod from the cradle. Whenever the barrel is removed, the recoil cylinder rear head should be removed in order to clean the rear interior of the recoil cylinder and to inspect for excess oil leakage. At this time the rear head should be thoroughly cleaned and the relief valve should be lubricated and checked for correct functioning. Special oil as issued by the Ordnance Department may be used in the recoil mechanism. A full reserve of oil for the recoil system amounts to approximately 1½ fills of the screw filler. In using the screw filler, care must be exercised to prevent crossing the threads or breaking off the end of the filler. The screw handle must be turned with both hands. The amount of oil reserve in the system is shown by the position of the oil index with reference to the front face of the recuperator cylinder front head, as follows:

(1) *No reserve.* The indicator is well into the recess. The piece must not be fired in this condition.

(2) *Full reserve.* The end of the indicator is even with the front face of the recuperator cylinder front head. The oil index does not of itself show when there is an excess of oil reserve, as the addition of excess oil does not move the index out beyond the front face of the recuperator cylinder front head. The piece must not be fired with an excess reserve (see **b** below).

b. Operations prior to firing. (1) Before firing, the reserve oil should be extracted until an insufficient reserve is indicated; then a full reserve should be established by inserting oil until the index is flush with the front face of the recuperator cylinder front head.

(2) The rear end of the recoil cylinder, the filling and drain plug hole, and the oil index recess should be examined for oil leakage. The presence of a few drops of oil at any of these places is to be expected, but if there is an undue leakage the piece must not be fired, and the condition should be reported to ordnance personnel.

c. Operations during firing. During firing, the recoil mechanism should be maintained at full reserve and the slides kept clean and properly lubricated. The chief of section should constantly verify the complete return of the piece to battery. Firing may be continued as long as the piece returns to battery sufficiently for the firing mechanism to actuate the trigger shaft. The chief of section should observe constantly the behavior of the recoil mechanism in firing, and take action in the case of malfunctioning as is indicated below:

MALFUNCTION	CAUSE	CORRECTION
(1) Oil index projects less than the required distance.	(1) (a) Loss of reserve oil. (b) Loss of gas pressure either through the re-	(1) (a) Drain remainder of the oil reserve and refill to normal. (b) Gas escaping

MALFUNCTION	CAUSE	CORRECTION
	cuperator cylinder rear head or past the floating piston.	by the floating piston is indicated by an emulsified condition of the reserve oil drained off. If in reestablishing the oil reserve, the oil index does not move out and the oil screw filler works easily, the gas pressure has been lost. Substantiate this by an attempt to drain the mechanism. Oil will not spurt out unless some pressure is present. Report the malfunctioning to ordnance personnel for repair.
(2) Oil index remains stationary when the reserve is pumped in against evident pressure.	(2) The packing is too tight, or the index is broken, or the index is locked by some foreign substance.	(2) Withdraw all reserve oil, then insert approximately one and one-half fills of the oil screw filler. Tap the oil index lightly as oil is being added. If it still fails to function, report the mal-

MALFUNCTION	CAUSE	CORRECTION
		functioning to ordnance personnel for repair.
(3) Failure of howitzer to return to battery.	(3) (a) Insufficient oil reserve. (b) Dirt or obstruction on the slides. (c) Relief valve in recoil cylinder rear head not functioning. (d) Low nitrogen pressure, excessive internal friction; damaged slides, piston rod or piston.	(3) (a) Withdraw the reserve oil and establish a new full reserve. (b) Clean and lubricate the slides. (c) Remove recoil cylinder rear head and clean. If malfunction continues, report the malfunctioning to ordnance personnel for repair. (d) Report malfunction to ordnance personnel for repair.
(4) Return of howitzer to battery with a shock.	(4) Too much oil reserve.	(4) Withdraw reserve oil until index is halfway in and, when mechanism has cooled, refill to a full reserve.

67. TUBE ASSEMBLY, BREECH MECHANISM, AND FIRING MECHANISM. a. Operations during firing. (1) During firing, all exposed bearing surfaces must be kept clean and covered with a thin film of lubricating oil.

(2) Whenever the rate of firing permits, swab the bore with clean water and a sponge.

(3) The chief of section constantly observes the functioning of the material. Causes and corrections of malfunctioning of the breech and firing mechanisms are given in TM 9-731 E.

b. Operations after firing. (1) As soon as possible after firing, disassemble the breechblock and firing mechanism, clean all parts, and oil lightly. Thoroughly clean the bore.

68. SIGHTING AND FIRE CONTROL EQUIPMENT. a. General. Special care is required to insure the positive and accurate functioning of the sighting and fire control mechanism. Exercise care to prevent denting the soft metal surfaces or scratching the glasses. Remove dirt from optical surfaces by brushing with a camel's-hair brush or lens paper. Remove oil or grease from glass by applying ethyl alcohol, or if alcohol is not available, by breathing on the glass then wiping lightly with lens paper or a clean soft cloth. Keep the unpainted steel surfaces covered with a light film of oil as prescribed in current TMs to prevent corrosion.

b. In general, the sights are correct:

(1) *Panoramic telescope.* (a) *In direction*, if the azimuth scale and micrometer read zero when the line of sighting is in a plane parallel to the vertical plane passing through the axis of the bore.

(b) *In elevation*, if, with the elevation indexes of the rotating head set at zero, the line of sighting is parallel to the axis of the bore.

(2) *Elbow telescope*, if the line of sighting through the zero range line is in a plane parallel to the horizontal plane passing through the axis of the bore.

(3) If there is no excessive lost motion between the sights and the sight mounts.

c. Testing equipment. Equipment used in testing sights consists of bore sights, a gunner's quadrant, and may include a test target. The target for sight adjustment may be a distant terrain object, 1500 or more yards away, or a test target for use in close proximity. If the vehicle is on uneven ground, cant the test target to correspond with the cant of the vehicle. Tests can be made without the bore sights by sighting through the firing pin recess or a cartridge case with the primer removed, using improvised cross hairs at the muzzle.

d. Test of the gunner's quadrant. To test the gunner's quadrant, set the scales at zero, place it on the quadrant seat of the howitzer, and level the quadrant bubble by means of the elevating handwheel. Then reverse the quadant on its seat. The bubble should center itself. If it does not, and the quadrant must be used, center the bubble by using the micrometer knob. Take one-half the resultant reading and apply it as a correction in the proper direction on all future settings. The quadrant should be adjusted at the earliest opportunity by ordnance personnel if the error exceeds 0.3 of a mil.

e. Verification and adjustment of telescope mount and panoramic telescope. Periodically, and whenever the mechanism is found to be out of adjustment, a detailed test and adjustment should be made. For this purpose it is desirable to have the motor carriage on an approximately level surface. It is assumed that all lost motion has been eliminated and that cross level bubbles are in adjustment. (For details see TM 9-731E.) Tests and adjustments are performed in sequence as follows:

(1) *Elevation indexes of telescope mount.* Lay the howitzer at zero elevation with the gunner's quadrant and center the cross and longitudinal level bubbles. The elevation indexes should coincide; if they do not, loosen the two screws in the adjustable index and slide it into

coincidence with the fixed index. Tighten the screws and recheck.

(2) *Deflection and range scales of panoramic telescope.* Sight on the test target or a distant terrain object. Insure that the elevation indexes of the telescope mount are in coincidence. Using the appropriate knobs of the telescope, place the center vertical cross hair and zero horizontal cross hair of the telescope reticle on the proper portion of the test target. If the azimuth and elevation scales and their respective micrometer scales do not indicate zero deflection and elevation, adjust in the following manner:

(a) Turn azimuth worm knob until center vertical cross hair is on the appropriate line of the test target. Loosen screws in the azimuth micrometer knob and, while holding the azimuth worm knob, slip index until the zero is opposite the index. If the zero of the azimuth scale is not opposite the azimuth scale index, adjust the azimuth micrometer index to read zero at the same time that the azimuth scale index reads zero. Loosen the tangent locking screws at the front of the telescope socket and adjust the tangent screws until the vertical cross hair is on the appropriate line of the test target. If the error is too great to be corrected in this manner, turn the azimuth micrometer knob until the vertical cross hair is on the target line. Then adjust the azimuth and micrometer scale indexes to read zero. The main azimuth scale may be adjusted by loosening the four screws on the collar above the scale and slipping the scale until it reads zero. Tighten the screws and recheck.

(b) *Elevation adjustment.* Turn the elevating knob of the telescope until the zero horizontal cross hair of the reticle corresponds with the appropriate mark of the testing target. Loosen screws in end of knob, and, holding the knob, slip elevation micrometer scale until the zero graduation lines up with its index; then tighten screws and recheck the setting.

e. Verification and adjustment of range quadrant.

(1) Move range drum assembly indexes into coincidence. The elevation micrometer scale and the elevation scale should read zero. The range drum will indicate a reading slightly greater than zero if properly assembled.

(2) With the range drum properly assembled, if the elevation micrometer scale does not read zero, it is adjusted in the following manner:

(*a*) Loosen the three screws in micrometer knob.

(*b*) Without moving the knob, slide the zero of micrometer scale into coincidence with the index.

(*c*) Tighten and recheck.

(3) With the elevation micrometer scale in adjustment, if the elevation scale does not indicate zero, it is adjusted in the following manner:

(*a*) Loosen the two screws in index. Move index opposite the zero graduation.

(*b*) Tighten screws and recheck.

(4) With the elevation micrometer scale and the elevation scale in adjustment and set at zero, the axis of the bore is accurately leveled and the trunnions are approximately leveled, the angle of site is set at 300, and the cross level bubble is centered. If the longitudinal level bubble is not centered, the range quadrant is adjusted in the following manner:

(*a*) Center longitudinal bubble by means of angle-of-site knob.

(*b*) While holding the angle-of-site knob, loosen clamping screw in center of knob and slip angle-of-site micrometer to indicate zero.

(*c*) If necessary, loosen the two screws which secure the angle-of-site scale and shift the scale so that the

"3" graduation registers at the index line. Tighten the screws to secure the scale in this position.

(d) Recheck. Further adjustment, if required, is to be performed only by authorized ordnance personnel.

f. Verification and adjustment of elbow telescope and mount. With the axis of the trunnions approximately leveled, boresight on the test target. If the "N" cross hair is not on its line of the test target, put it on the following manner:

(1) Loosen the elevation clamping screw. Loosen the elevation adjusting screw lock nuts. Bring the center of the "N" cross hair to bear exactly on the horizontal line of the target by means of the two elevated adjusting screws.

(2) Tighten the lock nuts and note whether tightening has shifted the setting. Tighten the elevation clamping screw.

(3) The dot in the center of the "N" cross hair may be brought into coincidence with the vertical cross line of the test target by means of the azimuth adjusting screws.

Section X

DESTRUCTION OF EQUIPMENT

69. DESTRUCTION OF EQUIPMENT. a. General principles. (1) Situations may arise when materiel must be destroyed to prevent—

(a) Its capture intact by the enemy.

(b) Its use by the enemy, if captured, against our own or allied troops.

(2) The principles to be followed are—

(a) Methods for the destruction of materiel subject to capture or abandonment in the combat zone must be adequate, uniform and easily followed in the field.

(b) Destruction must be as complete as available time, equipment, and personnel will permit. If thorough destruction of all parts cannot be completed, the most important features of the materiel should be destroyed. *The same essential parts must be destroyed on all like units to prevent the enemy's constructing one complete unit from several damaged ones by "cannibalism."*

(c) The destruction of materiel, subject to capture or abandonment in the combat zone, *will be undertaken only when in the judgment of the military commander concerned such action is necessary. The destruction of materiel is a command decision to be implemented only on authority delegated by the division or higher commander.*

(3) Crews will be trained in the prescribed methods of

destruction. *Training will not involve the actual destruction of materiel.*

b. Methods. (1) The methods below are given in order of priority shown. Adhere to the sequence for each method.

(2) Certain methods require special tools and equipment such as TNT and incendiary grenades, which may not be items of issue normally. The issue of such special tools and materiel, the vehicles for which issued, and the conditions under which destruction will be effected *are command decisions in each case,* according to the tactical situation.

70. DESTRUCTION OF THE HOWITZER. a. Sights. Detach all optical sights. *If evacuation is possible, carry the sights;* if evacuation is not possible, thoroughly smash the sights.

b. Method No. 1. (1) Open drain plugs on recoil mechanism, allowing recoil fluid to drain. *It is not necessary to wait for the recoil fluid to drain completely before firing the howitzer in* (d) *below.*

(2) Place an *armed (safety pin removed)* M9A1 antitank grenade, HE, or armed (safety pin removed) M6 antitank rocket in the tube about 6 inches in front of, and with the ogive nose end toward, the HE shell in (3) below.

(3) Insert an HE shell with propelling charge into the cannon and close the breech. Base detonating HE shell cannot be used in this method.

(4) Fire the piece with a lanyard from a distance of at least 100 yards. The person firing should be under cover to the rear of the piece and approximately 20° off the line of fire. Elapsed time: Approximately 2 to 3 minutes.

c. Method No. 2. (1) See **b** (1) above.

(2) Fire an HE round, assembled with a point deton-

ating fuze, against a similar round jammed in the muzzle. Take same precautions as in **b** (4) above.

d. Method No. 3. Insert 3 to 5 TNT blocks in the bore near the muzzle, and 8 to 10 in the chamber of the howitzer. Close the breechblock as far as possible without damaging the safety fuze. Plug the muzzle tightly with earth to a distance of approximately 3 calibers (12 inches). Detonate the TNT charges simultaneously.

e. Method No. 4. With another gun fire at the tube of the howitzer until it is damaged beyond usefulness.

f. Method No. 5. Insert four unfuzed M14 incendiary grenades end to end midway in the tube at 0° elevation. Ignite these four grenades by a fifth equipped with a 15-second Bickford fuze. The metal from the grenades will fuze with the tube and fill the grooves. Elapsed time: 2 to 3 minutes.

71. DESTRUCTION OF THE CALIBER .50 MACHINE GUN. a. Method No. 1. Field strip. Use barrel as a sledge. Raise cover; lay bolt in feedway; lower cover on bolt, smash down over bolt. Deform back plate. Wedge buffer into rear of casing allowing depressors to protrude; break off depressors by striking with barrel. Lay barrel extension on its side. Hold down with one foot, break off the shank. Deform casing by striking side plates just back of the feedway. Elapsed time: 3½ minutes.

b. Method No. 2. Insert bullet point of complete round into muzzle and bend case slightly, distending mouth of case to permit pulling of bullet. Spill powder from case, retaining sufficient powder to cover the bottom of case to a depth of approximately ⅛ inch. Reinsert pulled bullet, *point first,* back into the case mouth. Chamber and fire this round with the reduced charge; the bullet will stick in the bore. Chamber one complete round, lay weapon on ground, and fire with a lanyard at least 30 feet long. Use the best available cover as this means of destruction may be dangerous to the person destroying the weapon. Elapsed time: 2 to 3 minutes.

c. **Machine-gun tripod mount, caliber .50 M3.** Use machine-gun barrel as a sledge. Deform pintle yoke. Deform traversing dial. Fold rear legs and deform so as to prevent unfolding. Remove front leg and knock off yoke. Extend elevating screw and bend screw by striking with barrel. Turn mount over; deform head assembly and knock off dial locking screw and pintle lock. Elapsed time: 3 minutes.

72. DESTRUCTION OF THE M7 MOTOR CARRIAGE. a. This mount is destroyed the same as a medium tank, M4.

(1) *Method No. 1.* (a) Remove and empty the portable fire extinguishers. Puncture fuel tanks if practicable. Use fire of a cannon, or a fragmentation grenade for this purpose. Place TNT charges as follows:

   3 pounds between engine oil cooler and right
   fuel tank.
   2 pounds on the left side of the transmission as
   far forward as possible.

Insert tetryl non-electric caps with at least 5 feet of safety fuze in each charge. Ignite the fuzes and take cover. Elapsed time: 1 to 2 minutes if charges are prepared beforehand and carried in the vehicle.

(b) If sufficient time and materials are available, additional destruction of track-laying vehicles may be accomplished by placing a 2-pound TNT charge about the center of each track-laying assemblage. Detonate those charges in the same manner as the others.

(c) If charges are prepared beforehand and carried in the vehicle, *keep the caps and fuzes separated from the charges until used.*

(2) *Method No. 2.* Remove and empty the portable

fire extinguishers. Puncture fuel tanks. (See (1) (*a*) above). Fire on the vehicle using adjacent tanks, antitank or other artillery, or antitank rockets or grenades. Aim at the engine, suspension, and armament in the order named. If a good fire is started, the vehicle may be considered destroyed. Elapsed time: About 5 minutes per vehicle. Destroy the last remaining vehicle by the best means available.

(3) *Method No. 3.* Remove and empty the portable fire extinguishers. Puncture fuel tanks. Smash all vital elements (such as distributor, radiator, engine block, air and oil cleaners, generator, control levers, crankcase and transmission) with a heavy axe, pick, or sledge. Pour spare gasoline, oil, or distillate over entire unit and ignite.

c. (1) Whenever time and material are available, combine the vehicle destruction with the armament destruction previously described.

(2) If possible, detach and evacuate all machine guns mounted in the vehicle.

73. DESTRUCTION OF AMMUNITION. a. General.
(1) Time will not usually permit the destruction of all ammunition in forward combat zones.

(2) When sufficient time and materials are available, ammunition may be destroyed as indicated below. At least 30 to 60 minutes may be required to destroy adequately the ammunition carried by combat units.

(3) In general, the methods and safety precautions outlined in TM 9–1900, should be followed whenever possible.

b. Unpacked complete round ammunition. (1) Stack ammunition in small piles. (Small arms ammunition may be heaped.) Stack or pile most of the available gasoline in cans and drums around the ammunition. Throw onto the pile all available inflammable material such as rags, scrap wood, and brush. Pour the remaining available gasoline over the pile. Sufficient inflammable material must be used to insure a very hot fire. Ignite the gasoline and take cover.

(2) 105-mm howitzer ammunition can be destroyed by sympathetic detonation using TNT. Stack the ammunition in two stacks about 3 inches apart with fuzes in each stack toward each other. Place TNT charges between the stacks. Use 1 pound of TNT per 4 or 5 rounds of ammunition. Detonate all charges of TNT simultaneously from cover.

c. Packed complete round ammunition. (1) Stack the boxed or bundled ammunition in small piles. Cover with all available inflammable materials, such as rags, scrap wood, brush, and gasoline in drums or cans. Pour gasoline over the pile. Ignite the gasoline and take cover. (Small arms ammunition must be broken out of the boxes or cartons before burning.)

(2) The destruction of packed complete round ammunition by sympathetic detonation with TNT is not advocated for use in forward combat zone. To insure satisfactory destruction involves putting TNT in alternate cases or bundles of ammunition, a time-consuming job.

d. Miscellaneous. Grenades, antitank mines, and antitank rockets may be destroyed by the methods outlined above for complete rounds. The amount of TNT necessary to detonate these munitions is considered less than that required for detonating artillery shells. Fuzes, boosters, detonators, pyrotechnics, and similar material, should be destroyed by burning.

74. FIRE-CONTROL EQUIPMENT. a. All fire-control equipment, including optical sights and binoculars, is difficult to replace. It should be the last equipment to be destroyed, if there is any chance of personnel being able to evacuate. If evacuation of personnel is made, all possible items of fire-control equipment should be carried. If evacuation of personnel is not possible, fire-control equipment must be thoroughly destroyed as indicated in **b** and **c** below.

b. Firing tables, trajectory charts, slide rules, and similar items should be thoroughly burned.

c. All optical equipment that cannot be evacuated will be thoroughly smashed.

75. PNEUMATIC TIRES. a. General (1) Rubber is such a critical item that, whenever materiel is subject to capture or abandonment, an attempt to destroy pneumatic tires may always be made, *even if time will not permit destruction of the remainder of the vehicle.*

(2) With adequate planning and training, however, the destruction of tires may be accomplished in conjunction with destruction of the vehicle without increasing the time necessary.

b. Method No. 1. Ignite an M14 incendiary grenade under each tire.

c. Method No. 2. (1) Damage the tires with an axe, pick, or heavy machine gun fire (deflate them before doing this, if possible). Pour spare gasoline on tires, dousing each one, and ignite.

(2) When used in conjunction with wheeled vehicle destruction, the ensuing fire will adequately destroy the vehicle.

Index

	Paragraphs	Pages
Accuracy during laying	54	72
Aiming posts	55	72
Ammunition in direct fire	44	65
Ammunition:		
Care of	53	72
Preparation of, on ground	51	71
Repacking in containers	52	71
Ammunition trailer	14	24
Cannoneer:		
Duties of number 1	26, 27	44
Duties of number 2	28, 29	46
Duties of number 3	30, 31	47
Duties of number 4	32, 33	49, 51
Definitions	3	2
Destruction of equipment	69	87
Of ammunition	73	91
Of cal .50 machine gun	71	89
Of fire control equipment	74	92
Of the howitzer	70	88
Of the motor carriage M7	72	90
Of pneumatic tires	75	93
Dispersion	49	70
Displacement, correction of	56	73
Driver, duties of	22, 23	39
Fire, alert and system of	42	65
Ammunition	44	65
Commands	41	64
Direction of target	45	66
Dispersion	49	70

	Paragraphs	Pages
Lead	46	66
Range or elevation	47	67
Target identification	43	65
Fire, direct	34	55
Boresighting	36	57
Preparatory steps	35	55
Systems	37-40	59-62
Fire, indirect	19	34
Duties of chief of section	20, 21	34, 35
Firing, to cease	58	74
To suspend	59	74
Gunner, duties of	24, 25	40
Inspection and maintenance:		
Before operation	10	10
During operation	11	15
At halt	12	16
After operation	13	18
Lead	46	66
Materiel, maintenance of	62	76
Cleaning	64	77
Firing mechanisms	67	81
Inspections	63	77
Lubrication	65	78
Recoil mechanism	66	78
Sighting & fire control equipment	68	82
Tube assembly and breech	67	81
March order	18	31
Misfires	61	75
Night firing	50	70
Prepare for action	17	28

	Paragraphs	Pages
Purpose	1	1
Range	47	67
References	2	1
Section:		
Composition	4	4
Formation	5	4
Posts	6	6
To mount	7	6
To dismount	8	6
To post	9	9
Signals:		
Hand	16	26
Touch	15	26
Sighting, bore-	36	57
Target	43	65
Unload	60	74

©2013 Periscope Film LLC
All Rights Reserved
ISBN#978-1-940453-03-3
www.PeriscopeFilm.com

www.ingramcontent.com/pod-product-compliance
Lightning Source LLC
LaVergne TN
LVHW051847080426
835512LV00018B/3107